SpringerBriefs in Computer Science

For further volumes:
http://www.springer.com/series/10028

Nigel Boston

Applications of Algebra to Communications, Control, and Signal Processing

Nigel Boston
University of Wisconsin
Madison, WI, USA

ISSN 2191-5768 ISSN 2191-5776 (electronic)
ISBN 978-1-4614-3862-5 ISBN 978-1-4614-3863-2 (eBook)
DOI 10.1007/978-1-4614-3863-2
Springer New York Heidelberg Dordrecht London

Library of Congress Control Number: 2012935583

Printed on acid-free paper

Springer is part of Springer Science+Business Media (www.springer.com)

This book is dedicated to my wife, Susan Ann Boston.

Preface

Over the years, engineers and computer scientists have increasingly begun to use mathematical tools. Currently, it is standard for students in these fields to learn some complex analysis, probability, and statistics. There has, however, been a quietly growing introduction of tools from abstract algebra. Traditionally these have been used mostly in coding theory and cryptography, but there are emerging new areas of application for abstract algebra. For example, the first Applied Algebra Days conference, co-organized by the author and Shamgar Gurevich, took place in October, attracting speakers from Berkeley, Chicago, MIT, and other eminent institutions. That same month, the SIAM Activity Group in Algebraic Geometry held its first biannual conference. Both conferences covered very new approaches to applying algebra and both attracted researchers from mathematics, electrical engineering, and computer science.

This book is based on a 2004 course given by the author for graduate students from the Departments of Mathematics, Electrical Engineering, and Computer Science at the University of Wisconsin, but also incorporates a few more recent developments. It gathers into themes some of the main tools that are used today in applications and explains how they are used. This book should appeal to mathematicians wishing to see where what they know gets applied and to engineers and computer scientists wanting to learn more useful, modern tools.

Madison, WI

Nigel Boston
January 2012

Acknowledgements

Back in 1992, while visiting the Isaac Newton Institute in Cambridge, UK, which had just opened and where the library had very few books, I read the coding theory book by Van der Geer and Van Lint and suggested to my then student, Judy Walker, that she look into the subject. There began a journey into applications of algebra that was made easier by many helpful colleagues at the University of Illinois at Urbana-Champaign, most notably Dick Blahut and Ralf Koetter, but also Alex Vardy, Pierre Moulin, and Yi Ma. Since moving to the University of Wisconsin in 2002, several others have helped me in this quest, including Rob Nowak, Yu Hen Hu, Chuck Dyer, Steve Wright, and Bob Barmish. Outside of these institutions, other colleagues including Rob Calderbank, Gary McGuire, Judy Walker, Michael Gastpar, and Joachim Rosenthal, all of whom have worked on introducing deep algebraic tools into electrical engineering, have always been of the utmost help. I thank you all.

Contents

Acronyms

fpf	Fixed-point-free
LFSR	Linear feedback shift register
PSK	Phase-shift keying
SNR	Signal-to-noise ratio
STBC	Space-time block code
STTC	Space-time trellis code

Chapter 1
Introduction

Abstract This chapter gives an overview of the monograph.

The goal of this book is both to introduce algebraists to some novel applications of algebra in communications, control, and signal processing and also to teach engineers and computer scientists some advanced algebraic theory that can be employed in multiple settings. It begins with a discussion of the main tools of abstract algebra, namely groups and fields, and continues with a brief utilitarian guide to the main concepts of algebraic geometry, connecting them with the concept of field of functions. There emerge a couple of major ideas that appear repeatedly later on.

First, given a polynomial $p(x)$ with coefficients in a field F, we can define an addition and multiplication on the polynomials modulo $p(x)$. If $p(x)$ is irreducible, this is a way to produce fields that contain F and there are useful instances produced starting with F equal to the integers modulo a prime, the rational numbers, or even the rational functions defined with coefficients in some other field. There are applicable constructions even when $p(x)$ is reducible, such as $x^n - 1$ for various n, but these are not fields.

Second, if we do have one field E containing another F, then we can represent the elements of E by matrices with entries in F. For several applications we want certain sets of matrices and this is a method for producing them. This procedure can also be carried out when E behaves partially like a field except that it has noncommutative multiplication. Such objects are called division rings and no examples were known until Hamilton discovered the quaternions in 1843. His construction can be generalized to produce many examples of division rings, useful in applications. At the end of the chapter on algebraic tools, an example is given of how three-dimensional rotations can be conveniently described by quaternions. Moreover, one can list all finite groups of such rotations.

The following chapter gives a quick survey of older applications of algebra. These fall into two classes, namely to coding theory and to cryptography, which concern respectively the accurate and secure transmission of information. There are already many good texts in these areas and we simply give a brief introduction and illustrate how some of our thematic tools apply in these simple situations.

In coding theory, redundancy is exploited to help control and correct errors introduced by noise in the transmission channel. Mathematically, the goal is to find well-spaced vectors with entries in a finite field. One interesting class of codes is that of cyclic codes, and we see that their construction can be explained in terms of polynomials modulo $x^n - 1$. We also mention the most popular class of codes, the Reed-Solomon codes.

In public-key cryptography, security typically depends on our inability to solve some mathematical problem. With our improved ability to factor large integers, key-lengths in RSA have been increasing, and for handheld devices with constrained key-lengths, users have often turned to other cryptosystems. In particular, Elliptic Curve Cryptography is preferred in many scenarios. We touch on this and other cryptosystems using curves. Also, pseudorandom sequences produced by linear feedback shift registers can be used for private-key cryptography and it is explained how this is related to a construction of finite fields using specific primitive polynomials.

The next chapter turns to some modern aspects of coding theory. One development has been to improve upon point-to-point communication by using multiple antennas. In this case, it emerges that one wants well-spaced matrices instead of well-spaced vectors. There are two approaches to this. One uses collections of matrices that are closed under multiplication, in other words are finite groups of matrices. The other uses collections that are closed under addition. Such a set would have to be infinite, but suitable finite subsets work well. In 1998, Alamouti produced such a scheme, which, suitably interpreted, is simply the matrix representation of the quaternions. Decoding is accomplished by multiplicative inversion and this is why having all nonzero elements be invertible, as happens in a division ring, is important. There are very few division rings containing the reals but this is not a problem - in applications, we actually work with a finite modulation alphabet, which lives in a much smaller field, namely a cyclotomic extension of the rationals, and there are many useful division rings containing such a field.

There next comes a brief discussion of algebraic-geometry codes, which provide a powerful generalization of the Reed-Solomon codes. Good examples of these are constructed by finding curves defined over a finite field, where the curve has many points but small genus. Such curves have a second application. As noted by Niederreiter and Xing, they serve to construct the best known low discrepancy sequences. These are quasirandom sequences in large-dimensional unit cubes that can be used to give estimates for integrals over the cube with small, explicit error. In many applications, such as in the pricing of collections (tranches) of mortgages, these quasi-Monte Carlo methods are preferred to the traditional Monte Carlo methods which only promise heuristic error bounds.

The final chapter discusses applications of algebra to problems from signal processing, image processing, and control design. The first main concept here is that of the transfer function of a filter. This is a rational function over the reals, meaning that its numerator and denominator are polynomials with real coefficients. We give various examples where manipulating this transfer function leads to problems that can be addressed using techniques from algebraic geometry. This includes the famous Makhoul's Conjecture and Belgian Chocolate Problem. The latter is a plant stabilization problem in control design and it illustrates how the solution to certain optimization problems tend to be algebraic in nature. By identifying the algebraic properties this limiting solution must satisfy and thereby solving for it, one can perform an end run, circumventing the numerical methods normally used. Finally, a discussion of some algebraic methods that can be successfully employed in the classical face recognition problem is given. Here polynomial invariants of groups of rotations and translations are put to good use.

Chapter 2
Algebraic Tools

Abstract We introduce the algebraic tools that will be applied in subsequent chapters. This includes groups, fields, division rings, and varieties, most notably curves. The approach is utilitarian, giving the definitions but with the main focus being on examples.

2.1 Groups

Many mathematical sets of objects come with ways of combining their elements, for instance (1) real numbers have addition, (2) nonzero complex numbers have multiplication, and (3) permutations (shuffles) of a pack of cards have composition. These are all examples of groups.

Definition 2.1. A *group* is a set G with a rule for composing any two elements (denoted by juxtaposition) which is associative (i.e. $a(bc) = (ab)c$), has an identity element e (i.e. $ae = a = ea$), and in which every element a has an inverse a^{-1} (i.e. $aa^{-1} = e = a^{-1}a$).

Other examples we will use frequently include matrix groups under multiplication. For instance, (4) let $GL_2(\mathbf{R})$ denote the set of 2 by 2 matrices with entries in the real numbers, $\mathbf{R}$, and having nonzero determinant. As is shown in some linear algebra courses, this is a group under matrix multiplication, defined as follows:

$$\begin{pmatrix} a & b \\ c & d \end{pmatrix} \begin{pmatrix} e & f \\ g & h \end{pmatrix} = \begin{pmatrix} ae+bg & af+bh \\ ce+dg & cf+dh \end{pmatrix} \tag{2.1}$$

where $a,b,c,d,e,f,g,h \in \mathbf{R}$.

Some groups are also commutative (i.e. $ab = ba$) and are then called *abelian*. Sometimes, but not always, the operation is then written $+$ (compare (1) and (2) above) and instead of a^{-1}, we write $-a$. The groups in (3) and (4) fail commutativity and are called *nonabelian*. Generalizing (3), the set of permutations of N objects, usually denoted $1,2,\ldots,N$, is a group, called the *symmetric group* on N letters and denoted S_N. It is nonabelian if and only if N is greater than 2.

One way of obtaining more groups is to take certain subsets of known groups.

Definition 2.2. Given a group G, a *subgroup* H of G is a subset of G that is itself a group under the composition rule of G.

For example, let D_N be the group of symmetries of the regular N-gon. This consists of N rotations (which themselves form the cyclic group C_N) and N reflections. It is easily checked to form a group, called the *dihedral group* with $2N$ elements. Each of its elements permutes the N vertices of the N-gon, thus making D_N a subgroup of S_N. Alternatively, embedding the N-gon in the plane $\mathbf{R}^2$, these symmetries are given by two-by-two matrices, making D_N a subgroup of $GL_2(\mathbf{R})$. Among the useful theorems on subgroups is Lagrange's theorem [1], which says that if G is a finite group with n elements, then the number of elements of any subgroup H is a divisor of n.

Another subgroup of $G = GL_2(\mathbf{R})$ is:

$$H = \{\begin{pmatrix} x & y \\ -y & x \end{pmatrix} \mid x, y \in \mathbf{R}, (x, y) \neq (0,0)\} \tag{2.2}$$

To check that a subset is a subgroup, note that associativity comes for free from G. It remains to check that if $a, b \in H$, then $ab \in H$; that the identity $e \in H$; and that if $a \in H$, then $a^{-1} \in H$.

Example 2.1. The above H is a subgroup of $GL_2(\mathbf{R})$.

Proof. Since $\begin{pmatrix} x & y \\ -y & x \end{pmatrix}\begin{pmatrix} u & v \\ -v & u \end{pmatrix} = \begin{pmatrix} xu - yv & xv + yu \\ -xv - yu & xu - yv \end{pmatrix} \in H$, H is closed under multiplication. The identity $e = \begin{pmatrix} 1 & 0 \\ 0 & 1 \end{pmatrix} \in H$. If $a = \begin{pmatrix} x & y \\ -y & x \end{pmatrix}$, then let $\bar{a} = \begin{pmatrix} x & -y \\ y & x \end{pmatrix}$. Since $a\bar{a} = (x^2 + y^2)e$, it follows that every $a \in H$ has an inverse, namely $(1/(x^2 + y^2))\bar{a}$.

Note that the above formulas are reminiscent of those used to show that the nonzero complex numbers form a group under multiplication.

Definition 2.3. If G and H are groups and there is a one-to-one correspondence $\phi : G \to H$ compatible with the composition rules of G and H in the sense that $\phi(ab) = \phi(a)\phi(b)$ holds, then ϕ is called an *isomorphism* and we say that G and H are *isomorphic.*

Let G be the group of nonzero complex numbers with multiplication and H be the above group of matrices. The map defined by $\phi(x + iy) = \begin{pmatrix} x & y \\ -y & x \end{pmatrix}$, where $i = \sqrt{-1}$, is a one-to-one correspondence and if we multiply complex numbers and multiply the corresponding matrices, then the answers match, as noted above. Thus, ϕ is an isomorphism. (Behind this isomorphism is the fact that $\phi(x+iy)$ gives the matrix representing multiplication by $x+iy$ on the basis $[1,i]$ of $\mathbf{C}$ over $\mathbf{R}$.)

One of the reasons for introducing isomorphism is that we do not wish to differentiate, for instance, between the group of permutations of three apples and the group of permutations of three oranges. As far as group theory is concerned, these groups are indistinguishable.

The above example of a one-to-one correspondence extends neatly by sending the zero complex number to the zero matrix, in that it also sets up an isomorphism between the group of all complex numbers under addition and the group of all two by two matrices with real entries under addition. Many sets come with two composition rules like this and the next section introduces a class of such sets called fields.

2.2 Fields and Division Rings

Definition 2.4. A field F is a set with two composition rules, addition $+$ and multiplication (denoted by juxtaposition), that are both commutative and associative, have distinct identity elements (denoted 0 and 1 respectively), satisfy that every element a has an inverse under $+$ (denoted $-a$) and every $a \neq 0$ has an inverse under multiplication (denoted a^{-1}), and satisfy the distributive law for multiplication over $+$, namely $a(b+c) = ab+ac$ and $(a+b)c = ac+bc$.

In other words, F and $F - \{0\}$ are abelian groups under addition and multiplication respectively, plus distributivity holds. Familiar examples include the rational numbers $\mathbf{Q}$, the real numbers $\mathbf{R}$, and the complex numbers $\mathbf{C}$. If F is a field, we can always obtain a new field by letting $F(t)$ denote the set of all rational functions over F in t. This means quotients of a polynomial in t whose coefficients are in F by another nonzero such polynomial. Here $f_1(t)/g_1(t)$ is the same as $f_2(t)/g_2(t)$ whenever $f_1(t)g_2(t) = f_2(t)g_1(t)$ and, by factoring, we can write these quotients in lowest terms just as with rational numbers. Addition and multiplication are defined in the same way that we add and multiply fractions.

Another example of a field is the running example above, where F consists of all two by two matrices of the form $\begin{pmatrix} x & y \\ -y & x \end{pmatrix}$ where $x, y \in \mathbf{R}$. All that remains to be checked is that the distributive law holds. Extending the notion of isomorphism, we say that F is isomorphic to the field $\mathbf{C}$.

Definition 2.5. If E and F are fields and there is a one-to-one correspondence $\phi : E \to F$ which is compatible with both addition and multiplication, meaning that $\phi(a+b) = \phi(a)+\phi(b)$ and $\phi(ab) = \phi(a)\phi(b)$, then we call ϕ an isomorphism and say that E and F are isomorphic.

If F is a subset of a field E and is itself a field under the composition rules of E restricted to F (such as $\mathbf{Q}$ inside $\mathbf{R}$), we call F a subfield of E. If we weaken the requirements by allowing multiplication to be noncommutative, we obtain a kind of algebraic object that is becoming increasingly important in applications.

Definition 2.6. Let D be a set with two composition rules, addition and multiplication, such that D is an abelian group under addition, $D-\{0\}$ is a group under multiplication, and the distributive law holds. Then D is called a *division ring*.

For many years, no example of a division ring was known. In 1843, Hamilton discovered the quaternions, which are described as follows. Consider a four-dimensional vector space D over $\mathbf{R}$ with basis denoted $\mathbf{1},\mathbf{i},\mathbf{j},\mathbf{k}$. To multiply two elements of this space, it is enough to explain how to multiply basis elements and then extend by linearity. The element $\mathbf{1}$ is the identity for multiplication and the other rules are $\mathbf{i}^2=\mathbf{j}^2=\mathbf{k}^2=-\mathbf{1}, \mathbf{ij}=\mathbf{k}=-\mathbf{ji}, \mathbf{ki}=\mathbf{j}=-\mathbf{ik}, \mathbf{jk}=\mathbf{i}=-\mathbf{kj}$.

This is extended in the natural way. As an example, $(\mathbf{i}+\mathbf{j})+(\mathbf{2k}-\mathbf{i})=\mathbf{j}+\mathbf{2k}$ and $(\mathbf{i}+\mathbf{j})(\mathbf{2k}-\mathbf{i})=\mathbf{2ik}-\mathbf{i}^2+\mathbf{2jk}-\mathbf{ji}=-\mathbf{2j}+\mathbf{1}+\mathbf{2i}+\mathbf{k}=\mathbf{1}+\mathbf{2i}-\mathbf{2j}+\mathbf{k}$.

Theorem 2.1. *With these composition rules, D is a noncommutative division ring (called the ring of quaternions and usually denoted* $\mathbf{H}$*).*

Proof. Since D is a vector space, it is a group under $+$. By definition, $D-\{0\}$ is closed under multiplication, it has identity $\mathbf{1}$, and associativity and distributivity are easily checked. Since $\mathbf{i}$ and $\mathbf{j}$ do not commute, D is noncommutative. What about inverses?

Well, if α is an arbitrary element of D, say equal to $a\mathbf{1}+\mathrm{b}\mathbf{i}+\mathrm{c}\mathbf{j}+\mathrm{d}\mathbf{k}$ $(a,b,c,d \in \mathbf{R})$, then we define its conjugate $\overline{\alpha}$ to be $a\mathbf{1}-\mathrm{b}\mathbf{i}-\mathrm{c}\mathbf{j}-\mathrm{d}\mathbf{k}$ and see that $\alpha\overline{\alpha}=(a^2+b^2+c^2+d^2)\mathbf{1}$, because all other terms cancel in pairs. It follows that if $(a,b,c,d)\neq(0,0,0,0)$, then α^{-1} exists and equals $(1/(a^2+b^2+c^2+d^2))\overline{\alpha}$. Note the similarities between this and Example 2.1 above. This establishes that D is a division ring. In the above, $\sqrt{a^2+b^2+c^2+d^2}$ is called the *norm* of α and denoted $||\alpha||$.

Just as we found a matrix representation of $\mathbf{C}$ above, there is a useful matrix representation of $\mathbf{H}$.

Lemma 2.1. *Given* $\alpha=a\mathbf{1}+\mathrm{b}\mathbf{i}+\mathrm{c}\mathbf{j}+\mathrm{d}\mathbf{k}$*, let* $\phi(\alpha)=\begin{pmatrix} z & w \\ -\overline{w} & \overline{z}\end{pmatrix}$ *with complex entries* $z=a+ib, w=-c+id$*. Then* ϕ *is an isomorphism.*

Proof. One just checks that $\phi(\alpha\beta)=\phi(\alpha)\phi(\beta)$ and $\phi(\alpha+\beta)=\phi(\alpha)+\phi(\beta)$ for every $\alpha,\beta\in D$. It is also a one-to-one correspondence with the matrices of the given form, which ensures that these matrices form a division ring isomorphic to $\mathbf{H}$.

The representation of fields and division rings by matrices with entries in a subfield is one of the algebraic themes that recurs throughout this book. The other is a method that works in many contexts to construct fields and related structures. This is introduced in the next section.

2.3 Field Construction

The simplest example of field construction is the method of algebraic extension, described as follows. We need a preparatory lemma.

Lemma 2.2. *Let p be a prime. Given any integer n there is a unique integer $\overline{n} \in \{0,1,...,p-1\}$ such that $n-\overline{n}$ is divisible by p. This integer is called $n \pmod p$. Moreover, if n is not divisible by p, then there exist integers u, v such that $1 = up+vn$ and so $\overline{vn} = 1$.*

Proof. This is in many standard texts on abstract algebra, such as [1].

Now let $\mathbf{F_p} = \{0,1,...,\mathrm{p}-1\}$ (sometimes denoted $GF(p)$ by engineers) and define addition and multiplication modulo p as follows. For $a,b \in \mathbf{F_p}$, $a+b$ and ab may lie outside the range $[0, p-1]$ and so we replace them by $\overline{a+b}$ and $\overline{ab}$ in order to define addition and multiplication in $\mathbf{F_p}$.

Theorem 2.2. *$\mathbf{F_p}$ is a field with p elements.*

Proof. Addition and multiplication modulo p are easily shown to be commutative and associative. For example, if $a,b,c \in \mathbf{F_p}$, then $(a+b)+c = a+(b+c)$ in $\mathbf{Z}$ and each side $\pmod p$ is the same. They have identities 0 and 1 respectively. The additive inverse of n is $p-n$ ($n \neq 0$) and $-0 = 0$. What about multiplicative inverses?

Well, if $n \neq 0$, then as noted in the above lemma, there is an integer v such that $\overline{vn} = 1$, which says that $\overline{v}$ is the inverse of n. It remains to check distributivity, which follows since $a(b+c) = ab+ac$ in $\mathbf{Z}$ and so each $\pmod p$ is the same.

This set with its addition and multiplication is well-defined even if p is not prime, and is denoted $\mathbf{Z_p}$. It is, however, then not a field, since if p factors as $p = ab$ with $1 < a,b < p$, then in $\mathbf{Z_p}$, $ab = 0$. If $\mathbf{Z_p}$ were a field, we could then multiply by the multiplicative inverses of a and b to get $1 = 0$, a contradiction.

The above construction can now be generalized to produce new fields from old ones. Let F be a field and $p(x)$ a nonconstant polynomial with coefficients in F. We call $p(x)$ irreducible if it does not factor into a product of nonconstant polynomials. If $p(x)$ is quadratic or cubic, this is the same as saying it has no root in F, but this is not necessarily true for higher-degree polynomials. For example, x^4+4 has no roots in $\mathbf{Q}$, but factors as $(x^2+2x+2)(x^2-2x+2)$. It can be tricky to check whether high degree polynomials are irreducible, but there are computer algebra systems, such as MAGMA, that can handle this. These irreducible polynomials are the analogues of prime numbers and there is a lemma corresponding to Lemma 2.2, which we will exploit.

Lemma 2.3. *Let $p(x)$ be an irreducible polynomial of degree n. Given any polynomial $f(x)$ there is a unique polynomial $\overline{f}(x)$ of degree less than n such that $f(x) - \overline{f}(x)$ is divisible by $p(x)$. This is called $f(x) \pmod{p(x)}$. Moreover, if $f(x)$ is not divisible by $p(x)$, then there exist polynomials $u(x), v(x)$ such that $1 = u(x)p(x) + v(x)f(x)$ and so $\overline{v(x)f(x)} = 1$.*

Proof. Again, this is in any standard text on abstract algebra, such as [1].

Next, we define addition and multiplication modulo $p(x)$. Suppose $p(x)$ is irreducible of degree n and let E consist of all polynomials in x with coefficients in F and degree less than n:

$$E = \{a_0 + a_1x + \ldots + a_{n-1}x^{n-1} \mid a_i \in F\}$$

We will turn E into a field by suitably defining addition and multiplication. With addition there is no problem since the usual addition gives an answer in E. The product $f(x)$, however, of two polynomials in E can have degree greater than or equal to n and in that case we replace it by $\overline{f}(x)$. This defines a multiplication on E.

Theorem 2.3. *E is a field.*

Proof. The proof follows that of the previous theorem. As above, the main step is to note that if $f(x) \in E$ is not the zero polynomial, then there exists by lemma 2.3 a polynomial $v(x)$ such that $\overline{v}(x)$ is the inverse of $a(x)$.

This constructs a field E that contains F as a subfield by identifying F with the polynomials of degree zero in E. Notice that, ignoring the multiplication in E, E is a vector space over F of degree n, with basis $1, x, \ldots, x^{n-1}$. We now consider some examples of this construction.

2.3.1 Quadratic Fields

Let $F = \mathbf{Q}$ and $p(x) = x^2 - d$, where $d \in \mathbf{Z}$ is not a square. This makes $p(x)$ irreducible and then E consists of all polynomials $a + bx$ $(a, b \in \mathbf{Q})$, where, in E, $x^2 = d$ since their difference, $x^2 - d$, is clearly divisible by $p(x)$. This means that in E, x is a square root of d, which we denote $\sqrt{d}$, and we write that $E = \{a + b\sqrt{d} \mid a, b \in \mathbf{Q}\}$. This field is often denoted $\mathbf{Q}(\sqrt{\mathrm{d}})$. Examples like this, where $p(x)$ is irreducible of degree 2, are called *quadratic* extensions of F. For instance, the only quadratic extension of $\mathbf{R}$ is $\mathbf{C}$, constructed by taking any irreducible quadratic over $\mathbf{R}$ – conventionally $x^2 + 1$ is used, so that in E the basis elements are 1 and a square root i of -1.

We can take quadratic extensions of other fields. For example, if p is $3 \pmod 4$, then $x^2 + 1$ is irreducible over $\mathbf{F_p}$ (since its multiplicative group has order $p - 1$ and so by Lagrange's theorem has no element of order 4) and we thus obtain a field with p^2 elements. In fact, there exist irreducible quadratic polynomials over every $\mathbf{F_p}$. A *monic* polynomial is one with leading coefficient equal to 1.

Lemma 2.4. *For every prime p, there exist $(p^2 - p)/2 \geq 1$ irreducible monic quadratic polynomials over $\mathbf{F_p}$ and so there exists a finite field with p^2 elements.*

Proof. There are p^2 monic quadratic polynomials $a_0 + a_1x + x^2$. There are $p(p-1)/2$ polynomials of the form $(x+a)(x+b)$ with a, b distinct and p with $a = b$. This leaves $p^2 - p(p-1)/2 - p = (p^2 - p)/2$ irreducible monic quadratics.

In fact, for a given p, they all yield the same field up to isomorphism, denoted $\mathbf{F_{p^2}}$.

If $F = k(t)$, a field of rational functions, then so long as $f(t)$ is not a square, $x^2 - f(t)$ is irreducible and defines a quadratic extension of F, called a *hyperelliptic function field.*

2.3.2 Cyclotomic Fields

The other most important class of field extensions consists of the cyclotomic extensions. The polynomials $x^n - 1$ (for n greater than 1) are not irreducible. They always have factor $x - 1$ and if n is divisible by m, then $x^n - 1$ has factor $x^m - 1$. In a sense, these account for its entire factorization. For instance, if n is a prime power p^k, then there is a method (Eisenstein's criterion [1]) that tells us that $(x^{p^k} - 1)/(x^{p^{k-1}} - 1)$ is irreducible. This polynomial is denoted $\Phi_{p^k}(x)$. More generally, the *cyclotomic polynomial* $\Phi_n(x)$ is defined inductively by dividing $x^n - 1$ by $\prod \Phi_m(x)$ where the product is over all divisors m of n such that $m < n$. These polynomials always have integer coefficients.

Thus, $x^n - 1 = \prod \Phi_m(x)$ where the product is over all divisors m of n. For instance, $x^6 - 1 = (x-1)(x+1)(x^2+x+1)(x^2-x+1)$, where the factors are $\Phi_m(x)$ for the respective divisors $m = 1,2,3,6$ of 6. By induction, one then establishes that $\Phi_n(x)$ has degree $\phi(n)$ (Euler's totient function that enumerates how many integers d satisfy $1 \leq d \leq n$ and $\gcd(d,n) = 1$). The irreducibility of $\Phi_n(x)$ over $\mathbf{Q}$ for all n is a fairly deep result [1].

The corresponding extension E of $\mathbf{Q}$ constructed using $\Phi_n(x)$ is called the nth cyclotomic extension of $\mathbf{Q}$. In E, $x^n = 1$ since they differ by $x^n - 1$, which is divisible by $\Phi_n(x)$. Note that no smaller power m of x is 1 since $x^m - 1$ is not divisible by $\Phi_n(x)$. We say that E contains a primitive nth root x of 1, often denoted ζ_n, and write $E = \mathbf{Q}(\zeta_n)$.

2.4 General Field Extensions

When $F = \mathbf{Q}$, this method of construction (with a general polynomial) produces what are called number fields; when $F = k(t)$ (for some field k), it produces so-called function fields over k. The most useful case, however, is when $F = \mathbf{F_p}$. In this case an irreducible polynomial of degree n produces a field with p^n elements. Generalizing Lemma 2.4, such polynomials always exist and it turns out that for each prime power q there is, up to isomorphism, a unique field $\mathbf{F_q}$ with q elements. These are all the finite fields. They are useful in discrete applications.

For instance, taking $F = \mathbf{F_2}$ and $p(x) = x^8 + x^4 + x^3 + x + 1$ yields $\mathbf{F_{256}}$. We can put $a_0 + a_1x + \ldots + a_7x^7$ in correspondence with the binary string of length 8 (or byte)

$(a_0, a_1, ..., a_7)$ and the Advanced Encryption Standard in cryptography manipulates these bytes, as noted in the section on private-key cryptography later.

It turns out that, if $q = p^n$, the multiplicative group of $\mathbf{F_q}$ is always isomorphic to the cyclic group, $C_{|q|-1}$. This can be used to show that the cyclotomic polynomials over $\mathbf{F_p}$ are often not irreducible. For example, if p is $1 \pmod 4$, then the multiplicative group of $\mathbf{F_p}$ contains a primitive fourth root of 1 and so $\Phi_4(x) = x^2 + 1$ factors. If some cyclotomic polynomial is irreducible over $\mathbf{F_p}$, then it must construct a finite field, necessarily isomorphic to some $\mathbf{F_q}$.

We can always pick an irreducible polynomial $p(x)$ of degree n over $\mathbf{F_p}$ which yields a field E isomorphic to $\mathbf{F_q}$ with the additional property that its multiplicative group consists of all powers of x. For example, one can check that if $p = 2$ and $n = 4$, then $p(x) = x^4 + x + 1$ has this property - on the other hand, $p(x) = x^4 + x^3 + x^2 + x + 1$ is irreducible over $\mathbf{F_2}$ but that produces a field E in which $x^5 = 1$. In this field, there are only 5, not 15, distinct powers of x.

Cyclotomic extensions of $\mathbf{F_p}(\mathrm{t})$ are likewise of the form $\mathbf{F_q}(\mathrm{t})$ and are the somewhat uninteresting so-called constant field extensions.

We saw above some similarities between integers and polynomials. This extends much more deeply to an analogy between number fields and function fields over finite fields. These are the fields that our method constructs, starting with $F = \mathbf{Q}$ and $\mathbf{F_p}(\mathrm{t})$ respectively. The latter are related to curves, something we now explain briefly.

2.5 Curves and Function Fields

We will need some rudimentary definitions and theorems from algebraic geometry (see [2]).

Let k be a field and S be a finite set of polynomial equations in variables $x_1, ..., x_n$ with coefficients in k. A solution to the system S will be an ordered n-tuple in k^n satisfying all the equations in S. The set of all solutions is called a *variety*. We can consider the solutions of S over any field E containing k and denote that set by $V(E)$. An important example is the variety V where $k = \mathbf{Q}$ and S consists of one equation in two variables, $u^2 = f(t)$, with the coefficients of $f(t)$ in $\mathbf{Z}$. In the example given, we can even consider the equation over $\mathbf{F_p}$ by reducing the coefficients of $f(t)$ modulo p, but care must be taken since, for instance, $u^2 = f(t)$ and $u^2 = 4f(t)$ look much the same over $\mathbf{Q}$ but not over $\mathbf{F_2}$.

In the case that S consists of one equation in two variables, $g(t, u) = 0$ say, we call V a (planar) curve. We can associate to it a field E containing $k(t)$, via the earlier construction, so long as $g(t, u)$ is irreducible over $k(t)$. For instance, for the example just introduced, so long as $f(t)$ is not a square of a polynomial, then $E = \{a + bu \mid a, b \in k(t)\}$, where $u^2 = f(t)$ defines the multiplication in E. Notice that this is a quadratic extension of $k(t)$, analogous to a quadratic number field. If $f(t)$ has no repeated factor over any field containing k, we call V a *hyperelliptic curve*. This is more restrictive than just requiring that $f(t)$ is not a square; it ensures that the curve is smooth, meaning that there is no point on $V(E)$, for any field E containing k, at which the derivatives with respect to t and u both vanish.

2.5.1 Counting Points

Suppose that V is a curve over $\mathbf{F_p}$. Then, for each d, $V(\mathbf{F_{p^d}})$ is a finite set and we can ask how its size grows with d. For example, if V is the line given by $u = 0$, then $V(\mathbf{F_{p^d}})$ contains p^d elements. It is often better to work with what is called the projective completion of the variety. In this case, you get the projective line over $\mathbf{F_{p^d}}$. It contains one more element (the point at infinity) and so has $p^d + 1$ elements.

For a general smooth projective curve V over $\mathbf{F_p}$, there exist a positive integer g (the genus of V) and certain complex numbers $\alpha_1, \ldots, \alpha_g$ of absolute value $\sqrt{p}$ such that $V(\mathbf{F_{p^d}})$ contains exactly $p^d + 1 - \alpha_1^d - \overline{\alpha_1}^d - \ldots - \alpha_g^d - \overline{\alpha_g}^d$ points. This is one way to state the theorem of Hasse-Weil [13]. The projective line above is the case $g = 0$. It follows that the number of points in $V(\mathbf{F_{p^d}})$ lies between $p^d + 1 - 2g\sqrt{p}^d$ and $p^d + 1 + 2g\sqrt{p}^d$.

2.5.2 Elliptic Curves

One special case of hyperelliptic curve over $\mathbf{F_p}(\mathrm{p} > 2)$ is $y^2 = f(x)$ where $f(x)$ is a cubic polynomial with no repeated root in any field extension. Such curves, called *elliptic curves*, have the special property that their points form an abelian group, so long as we include an extra point, called the point at infinity ∞ (this gives the projective completion of the curve). The group law is given by saying that ∞ is the identity and that $P+Q+R$ equals the identity if and only if P, Q, R are collinear. The fact that $f(x)$ is cubic means that most lines intersect the curve in 3 points, and so the group law is well-defined. To add the same point to itself, we use the tangent to the curve at that point, and if the line joining two points to be added is vertical, we take their sum to be ∞. This does not yield a group law on other kinds of curves, but there is the notion of Jacobian of a curve of genus g, which is a g-dimensional variety that contains the curve and has a natural group law. For $g = 1$, the curve is elliptic and coincides with its Jacobian.

Since elliptic curves have genus $g = 1$, by Hasse-Weil, the number of points on an elliptic curve over $\mathbf{F_{p^d}}$ lies between $p^d + 1 - 2\sqrt{p}^d$ and $p^d + 1 + 2\sqrt{p}^d$. Note that knowing the number of points over $\mathbf{F_p}$ determines α_1 and hence the number of points over any $\mathbf{F_{p^d}}$.

2.6 Quaternions as Rotations

To give a taste of how abstract algebra can be useful in applications, we now discuss how the quaternions of unit norm, which we just met, can provide a compact way of describing three-dimensional rotations. This is a handy tool for people working in computer graphics and computer vision. First, recall that an isometry is a map from a space to itself that preserves distances.

Definition 2.7. The *orthogonal group* O_N is the set of all isometries of $\mathbf{R^N}$ that fix the origin. The *special orthogonal group* SO_N is the subgroup of isometries that are given by a matrix of determinant 1.

Every element of O_N (respectively SO_N) is a product of (respectively an even number of) reflections. A simple rotation is a product of two reflections and so lives in SO_N. In two dimensions we understand rotations about the origin well - they are just given by their angle of rotation.

$$SO_2 = \{ \begin{pmatrix} \cos\theta & -\sin\theta \\ \sin\theta & \cos\theta \end{pmatrix} \mid \theta \in [0, 2\pi) \} \tag{2.3}$$

Denote the given matrix by g_θ. The fact that rotations are simply described by their angle amounts mathematically to the following. As with the constructions above, one can talk of real numbers modulo $\mathbf{Z}$. Namely, let $G = [0, 1)$ with addition of $a, b \in G$ defined by taking the fractional part of $a + b$. This is a group and one checks that there is an isomorphism $\phi : SO_2 \to G$ given by $\phi(g_\theta) = \theta/(2\pi)$. This shows that SO_2 is an abelian group, because G is.

What about SO_3? In fact, this is related to the group of quaternions of norm one under multiplication. If β is such a quaternion, let $T_\beta : \mathbf{H} \to \mathbf{H}$ be given by $T_\beta(\alpha) = \overline{\beta}\alpha\beta$.

Lemma 2.5. *T_β acts trivially on the one-dimensional vector space with basis $\{\mathbf{1}\}$ and as an isometry with determinant 1 on the span of $\mathbf{i}, \mathbf{j}, \mathbf{k}$.*

Proof. $T_\beta(a\mathbf{1}) = \overline{\beta} a \beta = ||\beta||^2 a = a = a\mathbf{1}$. Also, $||T_\beta(\alpha)||^2 = \overline{\overline{\beta}\alpha\beta}\overline{\beta}\alpha\beta = \overline{\beta}\overline{\alpha}\beta\overline{\beta}\alpha\beta = \overline{\beta}\overline{\alpha}\alpha\beta = \overline{\beta}||\alpha||^2\beta = \overline{\beta}\beta||\alpha||^2 = ||\alpha||^2$. We leave it as an exercise to show that the determinant is 1 rather than -1.

In this way, $T_\beta \in SO_3$ and in fact every element of SO_3 is a simple rotation and is of this form. This representation of simple rotations is not quite unique since $T_\beta = T_{-\beta}$. The map $\pm\beta \mapsto T_\beta$ is, however, a one-to-one correspondence (indeed isomorphism) between quaternions of norm one, up to sign, and SO_3.

The finite subgroups of SO_2 are easy to classify. Namely, they are precisely the cyclic subgroups C_N (rotations of the regular N-gon into itself) introduced near the start of this chapter. Further investigation shows that we can classify all finite subgroups of SO_3, a result that is used in a variety of applications. Consider first the rotations of a regular tetrahedron. It is impossible in three-dimensional space to rotate it so that just two vertices are interchanged (a transposition) and so the group of symmetries is not all of S_4. In fact, the permutations of N symbols that are products of an even number of transpositions form a subgroup of S_N, called the *alternating group* on N letters and denoted A_N. It has half its size, i.e. $(N!)/2$ elements and A_4 and A_5 are isomorphic to the groups of symmetries of the regular tetrahedron and icosahedron respectively.

Theorem 2.4. *Every finite subgroup G of SO_3 is isomorphic to one of the following: C_N, D_N, A_4, S_4, or A_5.*

Proof. Here is an outline. Say $|G| = N$. Every nontrivial $g \in G$ is rotation about some line L, which intersects the unit sphere in $\mathbf{R}^3$ in two points $\pm p$. Gathering together the $g \in G$ that share the same p yields the stabilizer $G_p = \{g \in G \mid g(p) = p\}$. This is a subgroup of G_p and since it consists of rotations about a line, it is isomorphic to some cyclic group of size r_p. Counting corresponding pairs (g, p) gives $2(N-1)$ since each nontrivial g has two corresponding points $p, -p$. Alternatively, each p accounts for $r_p - 1$ nontrivial g. Therefore $2(N-1) = \sum_p (r_p - 1)$.

Let $O_p = \{g(p) \mid g \in G\}$. This is called the orbit of p under G. The map $g \mapsto g(p)$ sends $|G_p|$ elements of G to each element of O_p. It follows that $|G_p||O_p| = N$ (the orbit-stabilizer relation). Moreover, $h \in G_{g(p)} \iff g^{-1}hg \in G_p$ and so $|G_{g(p)}| = |G_p|$.

Thus, $2(N-1) = \sum n_i(r_i - 1)$, where the sum is over orbits O_i where $|O_i| = n_i$ and for $p \in O_i$, $|G_p| = r_i$. The orbit-stabilizer relation gives that $n_i r_i = N$. Then $2 - 2/N = \sum(1 - n_i/N) = \sum(1 - 1/r_i)$. This considerably limits the possible solutions. In particular, each term on the right side is at least $1/2$ and so there are at most three orbits.

If there is 1 orbit, then $2 - 2/N = 1 - 1/r_1$, which can never happen. If there are 2 orbits, then $2 - 2/N = 2 - 1/r_1 - 1/r_2$ and so $2/N = 1/r_1 + 1/r_2$. By Lagrange's theorem, N is divisible by r_1 and by r_2. Thus $r_1 = r_2 = N$ and G is isomorphic to C_N.

If there are 3 orbits, the possibilities are $r_1 = r_2, r_3 = N/2$, which lead to the dihedral group $G = D_{N/2}$. Else, the possibilities for r_1, r_2, r_3 are $2,3,3$ ($G = A_4$), $2,3,4$ ($G = S_4$), or $2,3,5$ ($G = A_5$).

References

1. Herstein,I.N. : Abstract Algebra. J. Wiley & Sons (1999)
2. Reid, M.: Undergraduate Algebraic Geometry, London Mathematical Society Student Texts 12, Cambridge University Press (1988)
3. Stichtenoth, H.: Algebraic Function Fields and Codes, Springer-Verlag (1993)

Chapter 3
Traditional Applications

Abstract The early applications of algebra to electrical engineering and computer science come via coding theory and cryptography. These are techniques used to ensure accuracy and security of transmitted information respectively. There are many books that cover these topics thoroughly. Our brief discussion is intended to provide background for the later chapters.

3.1 Coding Theory

In this section we summarize the main ideas of coding theory. A good general reference for those wishing to learn more is the book [1]. In traditional (channel) coding theory, we imagine transmitting vectors in a finite-dimensional vector space V over a finite field $\mathbf{F_q}$ and receiving slightly distorted vectors because of noise in the channel. The results are decoded to the nearest legitimate vector, as measured by Hamming distance, defined below. Unless otherwise specified, we will stick throughout with the most common case of binary vectors ($q = 2$).

Definition 3.1. Let V be a finite-dimensional vector space over $\mathbf{F_2}$. If $v, w \in V$, the *Hamming distance*, $d(v,w)$, between them is the number of coordinates in which v and w differ. The *Hamming weight* of a vector is its distance from the zero vector.

Lemma 3.1. *The function d is a metric on V. In other words, $d(v,w) = 0$ if and only if $v = w$; $d(v,w) = d(w,v)$; $d(u,w) \leq d(u,v) + d(v,w)$.*

Proof. The three properties follow immediately from the definition of d.

If we change the basis of V, we may change d, but we will think of V in terms of a standard basis $\{e_1,...,e_n\}$, meaning that $e_i \in V$ has a one in its ith coordinate, zero elsewhere. A code C is a subset of V whose elements are thought of as the codebook of legitimate vectors. In applications there will be some dictionary that translates between the actual codewords to be transmitted and vectors in C. Many of the most useful codes are not just subsets but in fact are subspaces (i.e. subgroups under addition) of V and we will also make this assumption throughout the book.

Definition 3.2. A (linear) code C is a subspace of an n-dimensional vector space V over $\mathbf{F_2}$. It is called an $[n,k,d]$-code if its dimension is k (and so it has 2^k elements) and if the minimum Hamming distance between two unequal vectors in C is d.

If v and w are a distance d apart, then $v-w$ and 0 are also d apart. Thus the parameter d is also the smallest weight of a nonzero vector in C. For a given n, there will be a trade-off between k and d, since the more vectors that are introduced, the more tightly they will be tend to be packed in. There follows an early example of a code whose vectors are optimally placed in that the balls of radius $\lfloor d/2 \rfloor$ around the code vectors fill up V and do not overlap.

Let V be seven-dimensional and take C to be the nullspace $\{x \in V \mid Hx^T = 0\}$ of the matrix $H = \begin{pmatrix} 0\,0\,0\,1\,1\,1\,1 \\ 0\,1\,1\,0\,0\,1\,1 \\ 1\,0\,1\,0\,1\,0\,1 \end{pmatrix}$. Such a matrix is called a *parity check matrix* of the code. The rows of H (the parity checks) are linearly independent, and so cut out a code C of dimension $7-3=4$. Its minimum distance is 3 and it is the $[7,4,3]$-code first found by Hamming. Note that the columns of H consist of all the nonzero vectors of $\mathbf{F_2^3}$. Doing the same with $\mathbf{F_2^m}$ produces a $[2^m-1, 2^m-m-1, 3]$-code, called the mth *Hamming code*.

In general, if we were to plot the rate $R = k/n$ against the relative minimum distance $\delta = d/n$ for every $[n,k,d]$-code, then because of the above trade-off we would obtain only points that are dense in the region below some curve $R = \alpha_2(\delta)$, whose equation is still not known. In practice we want codes with both R and δ large. The next section gives one method for constructing good codes.

3.2 Cyclic Codes

Let $p(x)$ be a polynomial with coefficients in $\mathbf{F_2}$. Recall that in the construction of field extensions E of $\mathbf{F_2}$, multiplication was defined modulo $p(x)$. This makes sense even if $p(x)$ is not irreducible – we just do not obtain a field, as was similarly observed with $\mathbf{Z_p}$ when p is not prime.

For instance, if $p(x) = x^7 - 1$, then E consists of the 2^7 polynomials of degree ≤ 6 with coefficients in $\mathbf{F_2}$ and $(x-1)(x^6+x^5+x^4+x^3+x^2+x+1) = x^7-1 = 0$ in E, which shows that $x-1$ can have no inverse (otherwise multiplying by it would give $x^6+x^5+x^4+x^3+x^2+x+1=0$).

Addition works just as before. The additive group of E above is isomorphic to a vector space of dimension seven over $\mathbf{F_2}$ via $a_0+a_1x+\ldots+a_6x^6 \mapsto (a_0,a_1,\ldots,a_6)$. and additive subgroups H of E thereby correspond to codes. Take, for instance, the elements of E of the form $(x^3+x+1)f(x)$ where $f(x)$ is a polynomial of degree ≤ 3. This is a 4-dimensional subspace and defines a $[7,4,3]$-code C - indeed, the Hamming code above! It does, however, have an additional property, which is that it is preserved under multiplication by x. This is explained in the next paragraph. Since $x^7 = 1$, multiplication by x translates into $(a_0, a_1, ..., a_6) \in C \Rightarrow (a_6, a_0, a_1, ..., a_5) \in C$, which is why these codes are called cyclic.

The reason that $x^3 + x + 1$ leads to the above property is that it is a factor of $x^7 - 1$. More generally, all codes with the property that $(a_0, a_1, ..., a_{n-1}) \in C \Rightarrow (a_{n-1}, a_0, a_1, ..., a_{n-2}) \in C$ arise by the following construction.

Definition 3.3. Let $E = \{a_0 + a_1 x + ... + a_{n-1} x^{n-1} \mid a_i \in \mathbf{F}_2\}$. Define addition and multiplication of elements of E by performing these modulo $p(x) = x^n - 1$. For any fixed factor $g(x)$ of $x^n - 1$ of degree $n - k$, let $H = \{g(x)f(x) \in E \mid \deg f(x) < k\}$. This subgroup H corresponds to an $[n, k, ?]$-code C. This is the *cyclic code* generated by $g(x)$.

Lemma 3.2. *If* $(a_0, ..., a_{n-1}) \in C$, *then* $(a_{n-1}, a_0, ..., a_{n-2}) \in C$.

Proof. Let $h(x) = (x^n - 1)/g(x)$. Then for any polynomial $f(x)$, $g(x)f(x) = g(x)\overline{f}(x)$ in E where $\overline{f}(x)$ is the unique polynomial of degree $< k$ such that $f(x) - \overline{f}(x)$ is divisible by $h(x)$, since their difference is divisible by $g(x)h(x) = x^n - 1$. Thus, multiplication by x maps H to itself.

An important family of such examples consists of the *quadratic residue codes.* Here n is a prime that is $\pm 1 \pmod 8$. Let K be the nth cyclotomic extension of $\mathbf{F}_2$. This means that K contains a primitive nth root α of 1. Then $x^n - 1$ factors completely into linear factors over K, as $x^n - 1 = \prod_{i=0}^{n-1}(x - \alpha^i) = (x - 1)\prod_{i \in S}(x - \alpha^i)\prod_{i \in N}(x - \alpha^i)$, where S and N are the set of squares and nonsquares respectively modulo n. We set $g(x)$ to be the first product and the cyclic code it produces is called the nth quadratic residue code. The condition on n in fact ensures that all the coefficients of $g(x)$ lie in $\mathbf{F}_2$.

For instance, if $n = 7$, $S = \{1, 2, 4\}$ and $\prod_{i \in S}(x - \alpha^i) = x^3 + x + 1$, and so, for $n = 7$, the quadratic residue code is the basic Hamming code. The quadratic residue code for $n = 23$ is also a famous code. It is a $[23, 11, 7]$-code, called the binary Golay code. It was used in the Voyager 1 and 2 missions, around about 1980.

The most commonly used codes are Reed-Solomon codes, and these work well, particularly when errors occur in bursts. They are cyclic codes but are defined over a more general finite field F than $\mathbf{F}_2$ such as $\mathbf{F}_{2^m}$ with e.g. $m = 8$. The idea is to pick n distinct nonzero elements $\alpha_1, ..., \alpha_n$ of F (it follows that $n \leq |F| - 1$) and to let C be the collection of vectors in F^n of the form $(f(\alpha_1), ..., f(\alpha_n))$ as f runs through all polynomials of degree $< k$ with coefficients in F. This produces an $[n, k, d]$-code over F with $d = n + 1 - k$, which meets the Singleton bound and so is as large as it can be.

We will generalize this construction later to algebraic geometric codes, but now we continue with a short introduction to some tools used in cryptography.

3.3 Linear Feedback Shift Registers

A linear feedback shift register (LFSR) produces a certain kind of pseudorandom binary sequence. It is constructed by a linear regression such as $x_n = x_{n-3} + x_{n-4}$. This is called a regression of degree four since each sequence element depends on the previous four. Starting with any initial sequence of length four, not all zero, this yields a longer sequence. For instance, $1,0,0,0,1,0,0,1,1,0,1,0,1,1,1,1,0,0,0,1,\ldots$, which repeats itself with period 15. Note that a sliding window of length four yields all nonzero vectors in $\mathbf{F}_2^4$ in this period and so the period is independent of which vector gives the initial sequence. This also shows that no regression of degree four has period greater than 15.

Likewise, for each N, we would like a linear regression of degree N that produces a sequence with period exactly $2^N - 1$ (the maximal possible) such that every nonzero binary vector of length N appears as a window of that length slides along the sequence. The theory of finite fields permits us to accomplish this.

Related to the above regression, consider $p(x) = x^4 + x + 1$. This is an irreducible polynomial over $\mathbf{F}_2$ that produces a field E with 16 elements. To each window (a,b,c,d) of the sequence, associate $a + dx + cx^2 + bx^3 \in E$. As the window slides, we obtain $(1,0,0,0),(0,0,0,1),(0,0,1,0),(0,1,0,0),(1,0,0,1),\ldots$, which correspond to $1, x, x^2, x^3, 1+x, \ldots$. Note that this last term is, in E, x^4. One calculates that sliding the window corresponds to multiplying by x in E.

Lemma 3.3. *Multiplication by x sends $(a,b,c,d) \mapsto (b,c,d,a+b)$.*

Proof. $x(a + dx + cx^2 + bx^3) = ax + dx^2 + cx^3 + bx^4 = ax + dx^2 + cx^3 + b(x+1) = b + (a+b)x + dx^2 + cx^3$

Thus, the period is the smallest power k of x such that $x^k = 1$ in E, which is 15. Notice that this does not work for all irreducible $p(x)$. If, instead, we used $p(x) = x^4 + x^3 + x^2 + x + 1$, corresponding to $x_n = x_{n-1} + x_{n-2} + x_{n-3} + x_{n-4}$, then we obtain sequences with period 5 since $x^5 = 1$ in E.

The key algebraic fact is that if E is any finite field, then the group $E - \{0\}$ under multiplication is isomorphic to the cyclic group $C_{|E|-1}$. This permits us to choose $p(x)$ not merely irreducible but also primitive, meaning that in the corresponding field E, x has maximal order $|E| - 1$ and thus the corresponding sequence has period $|E| - 1$.

In other words, this procedure does indeed yield, for every N, a periodic sequence of length $2^N - 1$, such that, in one period, a sliding window of length N will reveal every nonzero binary vector of length N.

3.4 Private-Key Cryptography

Pseudorandom sequences can be used for many purposes, one of which is to encode a binary sequence. Suppose we have a binary sequence (the message) $b_1, b_2, \ldots$ that we wish to transmit over a public channel. If the sender shares a secret in the form of a binary pseudorandom sequence (the key) $k_1, k_2, \ldots$ with the recipient, then the sender can simply transmit the "ciphertext" $b_1 + k_1, b_2 + k_2, \ldots$ instead. The recipient then subtracts the key in order to recover the message. This method is called the one-time pad and is perfectly secure so long as no information about the key is available.

In practice, however, it is prohibitively expensive (requiring couriers, for instance) to set up shared keys like this. The idea of pseudorandom sequences such as those from LFSRs is that a small amount of information (the $2N$ bits giving the regression and the initial sequence, or seed) leads to something with large period (length $2^N - 1$). The catch is that there exist efficient algorithms (such as Berlekamp-Massey) that will recover the seed from short portions (length $2N$) of the sequence. Modern-day commerce demands instant secure communication between parties who have never met, such as a customer buying online and the key distribution problem is solved as in the next section.

If two users do share a key (for instance, after a key exchange such as that explained below), they can use the Advanced Encryption Standard (AES), which is now the standard for private-key cryptography. Like its predecessor, the Data Encryption Standard (DES), it incorporates some nonlinear maps, but, whereas DES faced accusations that its nonlinear maps hid trapdoors that could be exploited by those in the know, the nonlinear map AES employs is much less artificial. It arises through identifying binary vectors of length eight (bytes) with elements of $\mathbf{F}_{256}$ and then is the map $u \mapsto u^{-1}$ $(u \neq 0)$, $0 \mapsto 0$, in this field.

3.5 Public-Key Cryptography

We give here a short overview of public-key cryptography. For more details, see for example the book by Blake, Seroussi, and Smart [2]. The main ideas of public-key cryptography were discovered in 1976 by Rivest, Shamir, and Adleman, who introduced the RSA system below. In 1997, it emerged that, during the period 1969-75, researchers at GCHQ, Cocks, Ellis, and Williamson, had also made this discovery. This was dedicated as an IEEE Milestone in 2010.

Suppose Alice wishes to send a message to Bob. If Bob had some locks made that were widely available and only he could open, then Alice could simply lock her message with Bob's lock and send it to him. Mathematically this is accomplished by a one-way function, standing for something that anyone can do, but is computationally hard to invert. The most common system is RSA, which depends on the fact that multiplying two large primes together is relatively easy, whereas factoring a product of two large primes is relatively hard. Recent advances in factoring, such as the number field sieve, have required that ever larger primes be used, and in constrained environments this can be impractical.

An alternative approach is Diffie-Hellman key exchange. Here we start with a finite group G and an element g of G, both of which can be made public. Alice and Bob choose integers a, b respectively and keep those private. Alice sends g^a to Bob and Bob sends g^b to Alice - these can be sent over a public channel. Then both Alice and Bob can compute $g^{ab} = (g^a)^b = (g^b)^a$ and use that as a shared key in the way described at the end of the last section.

An eavesdropper could break the system by computing a, given g and g^a (the discrete log problem in G), but this is often hard. The first kind of group used was the multiplicative group of a finite field. This cryptosystem turned out not to be as secure as a Diffie-Hellman system using the group of points on a carefully chosen elliptic curve over a finite field. This leads to elliptic curve cryptography (ECC), which is currently considered the best public-key system for many applications, so long as the curve is chosen carefully. More generally, people have investigated employing the group of points on Jacobians of other curves over finite fields. Hyperelliptic curve cryptography (HCC) arises this way. Currently, however, ECC is the most prevalent Diffie-Hellman system.

One of the most interesting parts of this theory is that there are certain elliptic curves that should be avoided or else the system's security can be weak. In particular, there can be problems depending on the number of points on the elliptic curve. For example, if the elliptic curve is supersingular over $\mathbf{F_q}$, which means that it has exactly $q+1$ points, then the Menezes-Okamato-Vanstone attack [2] is often effective. Much research has therefore gone into point-counting, beginning with the results of Chapter 2.

References

1. Blahut, R.M.: Algebraic Codes for Data Transmission. Cambridge University Press (2003)
2. Blake, I. , Seroussi, G., and Smart, N.: Elliptic Curves in Cryptography. London Mathematical Society Lecture Note Series (1999)

Chapter 4
Recent Applications to Communications

Abstract Modern-day communication often involves multiple transmitters and receivers. These are related by something called a channel matrix. The challenge of coding theory here becomes one of finding well-spaced matrices instead of well-spaced vectors, which can be accomplished through some group theory or division ring theory. The second topic in this chapter is quasirandom (so-called low discrepancy) sequences, which are good for applications in the theory of algebraic-geometry codes and also in mathematical finance.

4.1 Space-Time Codes

Space-time coding is a recently discovered technique for implementing antenna diversity. There are two branches, space-time block codes (STBC's) and space-time trellis codes (STTC's). Historically, STTC's came first, but Alamouti's scheme led research more in the direction of STBC's, which have simpler decoders. Alamouti's scheme is supported by third generation systems but has the drawback of being a single two-antenna system. We begin by describing it.

In 1998, Alamouti [2] proposed a coding scheme using two transmit antennas. The idea is to code both in time and space, hence the term *space-time coding*.

Suppose the ith antenna transmits complex number c_i at time 1. A receiver obtains $r_1 = h_1c_1 + h_2c_2 + \eta_1$, where h_1, h_2 describe the channel and η_1 is noise. Next, at time 2, the first antenna transmits $-\overline{c_2}$ and the second antenna $\overline{c_1}$ and $r_2 = h_1(-\overline{c_2}) + h_2\overline{c_1} + \eta_2$ is consequently received.

To perform maximum likelihood decoding, we would like to find the "codeword" (c_1, c_2) that minimizes $|\eta_1|^2 + |\eta_2|^2$. Write what we have so far as the following matrix equation:

$$\begin{pmatrix} r_1 \\ -\overline{r_2} \end{pmatrix} = \begin{pmatrix} h_1 & h_2 \\ -\overline{h_2} & \overline{h_1} \end{pmatrix} \begin{pmatrix} c_1 \\ c_2 \end{pmatrix} + \begin{pmatrix} \eta_1 \\ -\overline{\eta_2} \end{pmatrix} \tag{4.1}$$

Recalling the representation of the quaternions α by 2 by 2 matrices in Lemma 2.1, note that the matrix in the above equation is of that form. Its conjugate $\overline{\alpha}$ is then $\begin{pmatrix} \overline{h_1} & h_2 \\ \overline{h_2} & -h_1 \end{pmatrix}$ and multiplying by it through the equation allows us to invert and solve for (c_1, c_2):

$$\begin{pmatrix} \overline{h_1} & h_2 \\ \overline{h_2} & -h_1 \end{pmatrix} \begin{pmatrix} r_1 \\ -\overline{r_2} \end{pmatrix} = (|h_1|^2 + |h_2|^2) \begin{pmatrix} c_1 \\ c_2 \end{pmatrix} + \begin{pmatrix} \eta_1' \\ \eta_2' \end{pmatrix} \tag{4.2}$$

4.1.1 Rayleigh Flat-Fading Model

Suppose there are M transmitting and N receiving antennas, and that the channel is quasistatic in some period of T time blocks, at which points there is simultaneous transmission. Then the received signal is a T by N complex matrix

$$X = \sqrt{\frac{\rho T}{M}} HS + W$$

where ρ is the SNR (signal-to-noise ratio), S the signal (a T by M complex matrix), H the channel matrix (an M by N complex matrix giving the fades), and W the noise (a T by N matrix, entries independent complex Gaussian random variables, mean 0, variance 1). Note that S is actually a codeword. An information symbol vector will be translated into S, with encoding in both space (the M rows) and time (the T columns). Our aim is to ensure diversity at the transmitters.

In practice, H is unknown both at transmitter and receiver, because of rapid channel changes, limited system resources, and so on. The following method, found independently [4],[5], allows us to adapt the same approach as above to an unknown channel by using differential encoding.

Ignoring the normalizing factor above, the received signal X is given by a matrix equation $X = HS + W$, where H is the channel matrix, S the transmitted signal, and W is noise. The matrices S will be taken from some codebook. In differential encoding, we assume that the ith codeword $S_i = S_{i-1}V_i$, where the V_i are square $M \times M$ matrices.

Then $S_i = S_0 V_1 V_2 ... V_i$ and $X_i = HS_i + W_i = HS_{i-1}V_i + W_i = (X_{i-1} - W_{i-1})V_i + W_i = X_{i-1}V_i + W_i'$. Note that the channel matrix is eliminated from the calculation and, for maximum likelihood decoding, we just need to minimize $||X_i - X_{i-1}V_i||$ where $||.||$ is the Frobenius norm on matrices, i.e. $||(a_{ij}||^2 = \sum_{i,j} |a_{ij}|^2$.

The question now is how to pick a good constellation (codebook) of M by M matrices $C = \{V_1, V_2, ..., V_L\}$ and to define what "good" means. We will impose a power constraint, namely that all $V \in C$ have determinant 1. First, by the Chernoff bound, the probability of opting for V' when V is used is given [4]:

$$P(V, V') \leq (1/2) \prod_{m=1}^{M} (1 + (\rho^2/(1+2\rho))(1 - \delta_m(V^*V')^2)$$

where δ_m gives the mth singular value and $*$ means conjugate transpose. It is therefore dominated, for large SNR, by the term $(1/2)|\det(V-V')|^{1/M}$ (the $1/M$ normalizes over varying M and the $1/2$ normalizes it to $[0,1]$). We are therefore interested in "constellations" of $M \times M$ matrices $C := \{V_1, ..., V_L\}$ for which the *diversity product*

$$\zeta(C) := \min_{V \neq V' \in C}(1/2)|\det(V-V')|^{1/M}$$

and $|C|$ are both large. If $\zeta(C) \neq 0$, we say that C has full diversity. These also have the advantage that $HV_i \neq HV_j$ if $i \neq j$. Note that the *transmission rate* R is given by $\log_2(L)/M$. It turns out that there are two main methods for obtaining good constellations.

4.1.2 Closed under Multiplication

The best reference for this section is [12]. For $M = 1$ and given L, we obtain the largest $\zeta(C)(= 2\sin(\pi/L))$ by taking C to be the $\{e^{2\pi ik/L} | 0 \leq k \leq L-1\}$. Note that these form a group under multiplication.

The largest $\zeta(C)$ is unknown for larger M, except for small L. For instance, for $L = 2, C = \{I_M, -I_M\}$ gives $\zeta(C) = 1$. For $L = 3$, the largest $\zeta(C)$ is apparently $\sqrt{3}/2$ [14].

For M by M matrices we could take a phase modulation scheme consisting of diagonal matrices, i.e. for some choice of integers $r_1, ..., r_M$

$$C = \{ \begin{pmatrix} \exp(2\pi jr_1k/L) & 0 & ... & 0 \\ 0 & \exp(2\pi jr_2k/L) & ... & 0 \\ 0 & 0 & ... & 0 \\ 0 & 0 & ... & \exp(2\pi jr_Mk/L) \end{pmatrix} : k = 0, 1, ..., L-1\}$$

Another idea is to use orthogonal designs, which is to take C inside the group of 2 by 2 complex matrices of the form $\{\begin{pmatrix} a & b \\ -\bar{b} & \bar{a} \end{pmatrix} \mid |a|^2 + |b|^2 = 1\}$, which is the group of quaternions of norm 1 discussed in chapter 2.

Note that here $\det(V_i - V_j) = |a_i - a_j|^2 + |b_i - b_j|^2$, the usual Euclidean metric. Thus the problem is just to design spherical codes on the 3-sphere in real 4-space. (Note that $|a|^2 + |b|^2 = (\mathrm{Re}(a))^2 + (\mathrm{Im}(a))^2 + (\mathrm{Re}(b))^2 + (\mathrm{Im}(b))^2$. It turns out that these known approaches are special cases of more general constructions, as follows.

Let C be a finite group of M by M complex matrices under multiplication. There are at least three advantages of the constellation forming a group. First, in computing S_i, the product $V_1V_2...V_i$ can be computed inside the group easily, e.g. by look-up. Second, letting I_M denote the M by M identity matrix, $\det(V - V') = \det(V)\det(I_M - V^{-1}V') = \det(I_M - V^{-1}V')$, which leaves only $L-1$ cases to check in order to compute $\zeta(C)$. Finally, powerful mathematical tools, such as representation theory, character theory, and the classification of finite matrix groups can be brought to bear. Note that the result, Theorem 2.4, at the end of chapter 2 gives this classification in the 2×2 case.

Having full diversity then corresponds to satisfying $\det(I_M - V) \neq 0$ for all $V \neq I_M$ in C. In other words, no nonidentity element of C has 1 as an eigenvalue. Such groups are called *fixed-point-free*(fpf) and were classified by Zassenhaus in 1936 [16] (Hassibi et al. [12] noticed and corrected a few small errors).

Example 4.1. Let $\omega = e^{2\pi i/21}$, $A = \begin{pmatrix} \omega & 0 & 0 \\ 0 & \omega^4 & 0 \\ 0 & 0 & \omega^{16} \end{pmatrix}$, and $B = \begin{pmatrix} 0 & 1 & 0 \\ 0 & 0 & 1 \\ \omega^7 & 0 & 0 \end{pmatrix}$. Let $C = \{A^k B^l \mid 0 \leq k < 20, 0 \leq l < 3\}$. This is a nonabelian group of order 63 and $\zeta(C) = 0.385$. By way of contrast, the best 3-antenna phase modulation scheme with $|C| = 63$ has $\zeta(C) = 0.330$. This example was implemented by a wireless apparatus at Bell Labs, Murray Hill. The three transmit antennas were separated from the one receive antenna by 10 meters around a bend in a hallway lined with metal walls and equipment for a quasistatic scattering environment. It was practical, and simulations came out well.

One of the best schemes discovered to date is the following:

Example 4.2. Let $\omega = e^{2\pi i/5}$,

$$A = (1/\sqrt{5}) \begin{pmatrix} \omega^2 - \omega^3 & \omega - \omega^4 \\ \omega - \omega^4 & \omega^3 - \omega^2 \end{pmatrix}, B = (1/\sqrt{5}) \begin{pmatrix} \omega - \omega^2 & \omega^2 - 1 \\ 1 - \omega^3 & \omega^4 - \omega^3 \end{pmatrix}$$

The group C generated by A and B (meaning the smallest group containing them) has 120 elements and $\zeta(C) = 0.309$ (whereas orthogonal designs of about the same size do no better than 0.199 and 2-antenna phase modulation schemes of the same size no better than 0.135).

This is a famous group. It gives rise to the A_5 symmetry group of the icosahedron in theorem 2.4. It is also isomorphic to $SL_2(\mathbf{F_5})$, where $SL_2(F)$ denotes the group of 2 by 2 matrices of determinant 1 with entries in F. Shokrollahi [11] found a group of 8 by 8 complex matrices isomorphic to $SL_2(\mathbf{F_{17}})$, which is not fpf, but has relatively few matrices having 1 as an eigenvalue. This example turns out to be better at low SNR than the fpf groups.

Where do these examples come from algebraically? Besides Zassenhaus's work, which gives a list of candidate groups to consider, there are two other tools. First, given a finite group G, its representation theory gives all the ways in which it arises as a group of complex matrices. This is handily encapsulated in a square table called its character table and this was used by Shokrollahi to find his example above. Second, the finite subgroups of $SL_M(\mathbf{C})$ have been studied for many years. The theorem of 2.4 is used to classify them for $M = 2$ and they are now classified for all M up to 8. This was used [9] to study all finite constellations (fixed-point-free or not) closed under multiplication for small M.

4.1.3 Closed under Addition

Another approach to finding well-spaced matrices is to use constellations of matrices closed under addition. This would actually produce infinite codebooks and so in implementation we consider finite subsets of these. The idea is to generalize Alamouti's scheme, which can be interpreted in terms of Hamilton's quaternions. The bad news is that the only division ring with center the real numbers is $\mathbf{H}$ itself (and trivially $\mathbf{R}$) and so Alamouti's is one of a kind. The good news is that since the symbols we use (our modulation alphabet) is finite, we can employ a cyclotomic field in place of the reals, in which case many useful division rings exist.

First, note that if D is a division ring, then its center $Z(D) := \{x \in D \mid xy = yx \text{ for all } y \in D\}$ is a field K. Then D is a vector space over K and if finite-dimensional, then one can show that this dimension is a square d^2 and the maximal subfields of D all have dimension d over K. A case in point is $D = \mathbf{H}$, where $Z(D) = \mathbf{R}$, $d = 2$, and the maximal subfields of D are all isomorphic to $\mathbf{C}$. Since $\mathbf{C}$ is the only finite extension field of $\mathbf{R}$ besides $\mathbf{R}$, this leads to the shortage of useful division rings with center mentioned above.

We remedy this by constructing, for various m, a division ring with center K the mth cyclotomic field $\mathbf{Q}(\zeta_{\mathrm{m}})$. The idea is that we will use mth roots of 1 as symbols in a modulation alphabet. This is called the m-PSK (phase-shift keying) alphabet. For instance, in trellis coded modulation 8-PSK is used and if $m = 4$, this is called QPSK. In that case, $K = \mathbf{Q}(\mathrm{i})$, which also arises when the alphabet used is 16-QAM, which consists of the 16 points $a + bi$ with $a, b \in \{\pm 1, \pm 3\}$.

We have seen how the complex numbers can be represented as 2 by 2 matrices with entries in $\mathbf{R}$ and how the quaternions can be represented as 2 by 2 matrices with entries in $\mathbf{C}$. We now generalize these constructions.

Let D be a division ring containing K as a subfield such that D is an n-dimensional vector space over K. Pick a basis $\beta_1, \ldots, \beta_n$ for D over K so that $D = \{\lambda_1 \beta_1 + \ldots + \lambda_n \beta_n \mid \lambda_1, \ldots, \lambda_n \in K\}$. If $a \in D$, then the map $x \mapsto ax$ is a K-linear map of the vector space to itself and so with respect to the basis is represented by an n by n matrix, $\rho(a)$, with entries in K. The set of all such matrices forms a division ring isomorphic to D via the isomorphism ρ.

Example 4.3. Let $K = \mathbf{Q}(\mathrm{i})$ and D be the quadratic extension $K(\sqrt{i})$ of K (D is of course the cyclotomic field $\mathbf{Q}(\zeta_8)$, obtained by adding a primitive eighth root of 1 to $\mathbf{Q}$). A basis for D over K is given by $1, \sqrt{i}$. One calculates that if $x = a + b\sqrt{i}$, then $\rho(x) = \begin{pmatrix} a & ib \\ b & a \end{pmatrix}$, since $(a + b\sqrt{i})1$ gives the first column and $(a + b\sqrt{i})\sqrt{i} = ib + a\sqrt{i}$ the second.

Note that the determinant of $\rho(x)$ is $a^2 - ib^2$, which is 0 if and only if $a = b = 0$. We say the *strict rank* criterion holds. This means that not all nonzero matrices in the ambient vector space, not just the finite constellation, have full rank, which helps for fast decoding. This method allows us to construct some very good constellations.

Definition 4.1. Let F be a field and $\beta, \gamma \in F - \{0\}$. The corresponding *quaternion algebra* over F is $D = \{a + b\mathbf{i} + c\mathbf{j} + d\mathbf{k} \mid a, b, c, d \in F\}$ with addition defined coordinatewise and multiplication defined by $\mathbf{i}^2 = \beta, \mathbf{j}^2 = \gamma, \mathbf{ij} = \mathbf{k} = -\mathbf{ji}$.

This generalizes the example of $F = \mathbf{R}, \beta = \gamma = -1$, which produces $\mathbf{H}$. As with $\mathbf{H}$, we define the conjugate of $\alpha = a + b\mathbf{i} + c\mathbf{j} + d\mathbf{k}$ to be $\overline{\alpha} = a - b\mathbf{i} - c\mathbf{j} - d\mathbf{k}$ and the reduced norm $N(\alpha)$ to be $\alpha\overline{\alpha} = a^2 - \beta b^2 - \gamma c^2 + \beta\gamma d^2$. The argument that produces an inverse to $\alpha \in \mathbf{H}$ shows that D is a division ring if and only if for all $\alpha \in D - \{0\}$, $N(\alpha) \neq 0$.

Given such a division ring D, it has a subfield $K = F(\sqrt{\beta})$ over which D is a two-dimensional vector space. This allows the elements of D to be represented by 2 by 2 matrices over K, so that $\rho(\alpha) = \begin{pmatrix} a + b\sqrt{\beta} & c + d\beta \\ \gamma(c - d\beta) & a - b\beta \end{pmatrix}$. Note that $N(\alpha)$ is the determinant of $\rho(\alpha)$.

Example 4.4. Let $F = \mathbf{Q}(\mathrm{i}), \beta = \mathrm{i}, \gamma \in \mathrm{F}$. Then $K = \mathbf{Q}(\zeta_8)$, as in the previous example. It is clear that D is a division ring if and only if γ is not a norm from $\mathbf{Q}(\zeta_8)$ to $\mathbf{Q}(\mathrm{i})$.

Our constellation will consist of matrices of the form $\rho(\alpha)$ above, where $a, b, c, d \in F = \mathbf{Q}(\mathrm{i})$ are chosen in some finite subset of $\mathbf{Z}[\mathrm{i}] := \{\mathrm{r} + \mathrm{si} \in \mathbf{Q}(\mathrm{i}) \mid \mathrm{r}, \mathrm{s} \in \mathbf{Z}\}$.

Definition 4.2. A *perfect* space-time block code is one that has full rate, full diversity, uniform average transmitted energy per antenna, a nonvanishing minimum determinant for increasing spectral efficiency, and good shaping.

These are all the desirable properties we want of a space-time code. For example, the *golden code* below was discovered independently by several people.

Example 4.5. (The Golden Code) Let $F = \mathbf{Q}(\mathrm{i}), \beta = \mathrm{i}, \gamma = 5$. Let $\theta = (1 + \sqrt{5})/2, \overline{\theta} = (1 - \sqrt{5})/2$ and $\alpha = 1 + i - i\theta, \overline{\alpha} = 1 + i - i\overline{\theta}$. The golden code consists of matrices of the form $(1/\sqrt{5}) \begin{pmatrix} \alpha(a + b\theta) & \alpha(c + d\theta) \\ i\overline{\alpha}(a + b\overline{\theta}) & \overline{\alpha}(c + d\overline{\theta}) \end{pmatrix}$.

This is an example of a perfect space-time code, meaning full rate, full diversity, nonvanishing minimum determinant for increased spectral efficiency, and good shaping.

4.2 Network Coding

A more recent development has been that of network coding, where we find use for some tools from algebraic geometry. Imagine if Alice A and Bob B each want to send a message to the other. They try broadcasting, but cannot reach far enough. Imagine further that between them is person C, who can hear both and broadcast to both.

One thing that C can do is simply to alternate sending out the message of one, then the other, but by an elementary idea they can convey the information twice as fast. Namely, have C broadcast the sum of the two messages (added modulo 2, i.e. XOR, if these are in binary). If the message of A is a and of B is b, then both receive $a+b$ and can subtract (which is the same as add modulo 2) to recover the other's information.

The above is equivalent to the butterfly network that was introduced by Ahlswede, Cai, Li, and Yeung [1]. It is the basic example of network coding, where, instead of just relaying packets of information one at a time, the nodes of a network combine packets algebraically and thereby increase the throughput.

We represent a network as a directed graph with nodes (vertices) V, directed edges E linking certain pairs of nodes, and capacities C giving maximum information flow along each edge. If we want the maximum possible throughput from one node (a source) to another (a sink), there is a graph-theoretical bound. The point is that using traditional routing ideas often does not attain maximum throughput, but by allowing nodes to perform algebraic (linear) operations on the packets there are many scenarios in which it is attained.

In linear network coding, there are source nodes and sink nodes. Each intermediate node computes a new packet, which is a linear combination of the ones it receives, with coefficients in some finite field $\mathbf{F_q}$ (typically, q is a large power of 2). In this way, the network coding problem amounts to finding a transfer matrix M, entries in $\mathbf{F_q}$, that satisfies $z = xM$ for a given input vector x of information to be sent by the source nodes and output vector z to be received at the sink nodes.

By inserting variables for the entries of M, this condition translates into equations connecting those variables, such as that certain submatrices should be invertible, so that the information can be recovered at the sinks, just as from $a+b$ and a above, Alice could recover both a and b. These equations define a variety and the theorem of Koetter and Médard [6] deduces that the network coding problem is solvable if and only if that variety has at least one point (which then provides the coding scheme).

4.3 Low Discrepancy Sequences

For more on this topic, see Niederreiter's book [7].

In many applications, we wish to estimate the integral of a function over the unit cube in s dimensions, I^s. For instance, banks trade collections of mortgages called tranches and need to estimate the expected value of the sum of present values of future values for each tranche. There are proprietary formulae based on human behavior that give this value as an integral over I^{360}, since a mortgage typically runs over 360 months [10].

A Monte Carlo method would pick N random points $x_1, \ldots, x_N \in I^s$ and then approximate $\int_{I^s} f(u)du$ by $(1/N)\sum_{i=1}^{N} f(x_i)$. The problems with this are twofold. First, truly random points tend to cluster in places and second, the error bounds in this estimate are heuristic rather than deterministic. For these reasons, starting with Paskov and Traub in 1992 [10], some people in finance preferred to use quasirandom points to obtain a deterministic bound. These are called quasi-Monte Carlo methods. The next question then is to ask what properties of the points chosen are desirable. It turns out that their measure of goodness is star discrepancy.

Definition 4.3. Let $S = \{x_0, ..., x_{N-1}\} \subseteq I^s$. The *star discrepancy* of S is $D_N^*(S) = \sup_J |A(J,S)/N - Vol(J)|$, where the sup is over all J of the form $\prod_{i=1}^s [0,u_i)$, where $A(J,S) = |\{i \mid x_i \in J\}|$.

If S is an infinite sequence $x_0, x_1, ...$ in I^s, then $D_N^*(S)$ is defined by using its first N terms.

Theorem 4.1. *(Koksma-Hlawka) If f is a real-valued function defined on I^s and has bounded variation and if $S = \{x_0, ..., x_{N-1}\}$, then*

$$|(1/N) \sum_{i=0}^{N-1} f(x_i) - \int_{I^s} f(u)du| \leq V(f) D_N^*(S)$$

We are therefore interested in sequences with small star discrepancies.

Definition 4.4. The sequence $x_0, x_1, ...$ of points in I^s is called a *low discrepancy sequence* if there is a constant C depending on s and S such that $D_N^*(S) \leq CN^{-1}(\log N)^s$.

An example of this is the s-dimensional *Halton sequence*. Let $p_1, ..., p_s$ be the first s prime numbers. Writing each positive integer k to base p_j as $\sum a_{ij} p_j^i$, we define $\phi_{p_j}(k) = \sum a_{ij} p_j^{-i-1}$ and set $x_k = (\phi_{p_1}(k), ..., \phi_{p_s}(k)) \in I^s$. Halton proved in 1960 that this is a low discrepancy sequence.

Unfortunately, for large s, for these Halton sequences, the constant C in the definition above turns out to be large. The goal is to see how small we can make C through judicious choice of sequence.

There are certain kinds of low discrepancy sequences called digital (t,s)-sequences in base b. For these, $D_N^*(S)$ is low discrepancy with the constant C a complicated function of s times b^t. The smaller the star discrepancy is, the better the bound, and so for a given s we want t as small as possible. Niederreiter and Xing [8] used the theory of curves over finite fields $\mathbf{F_q}$ to produce (t,s)-sequences in base q with t the genus of a curve over $\mathbf{F_q}$ with at least $s+1$ points.

We therefore want curves over $\mathbf{F_q}$ with small genus and lots of points. According to the theorem of Hasse-Weil (section 2), a curve over $\mathbf{F_q}$ of genus g has at most $q+1+2g\sqrt{q}$ points. As noted in the second chapter, curves over $\mathbf{F_q}$ correspond to finite extensions of $\mathbf{F_q}(\mathrm{t})$ and deep methods of algebraic number theory are used to construct the Niederreiter-Xing examples.

In particular, they use Class Field Theory, which was the major topic of research for number theorists in the first half of the twentieth century. It allows one to describe certain extensions of a number field or function field F in terms of the arithmetic of F. For example, so long as F is not an extension of $\mathbf{F_2}(\mathrm{t})$, the quadratic extensions of F correspond to nonzero elements of F and two such give the same extension if and only if their ratio is a square in F. The extensions that can be described also include but generalize the cyclotomic extensions of F.

There are companies that have proprietary examples of good sequences produced from Class Field Theory. It is amazing that this advanced graduate algebra topic is useful to Wall Street companies.

4.3.1 Algebraic-Geometry Codes

A generalization of Reed-Solomon codes was discovered by Goppa in 1980 [3]. For more details on this section, see Stichtenoth's book [13].

Suppose C is a curve over $\mathbf{F_q}$. A divisor on C is a formal sum with integer coefficients of points on C. In other words, divisor $D = \sum_{P \in C} \lambda_P P$, where the $\lambda_P \in \mathbf{Z}$. It is called positive ($D \geq 0$) if all the $\lambda_P \geq 0$. Its degree is defined to be $\sum \lambda_P$. Since every nonzero rational function f on C has finitely many zeros and poles, associated to f is the divisor $\operatorname{div}(f) = \sum_{P \in C} v_P(f) P$, where $v_P(f)$ is the order of the zero of f at P or minus the order of the pole of f at P. This has degree zero. The space $\mathrm{L}(D) = \{f \neq 0 \mid \operatorname{div}(f) + D \geq 0\} \cup \{0\}$ is a finite-dimensional $\mathbf{F_q}$-vector space whose dimension is given by the Riemann-Roch theorem.

The idea behind Reed-Solomon codes was to evaluate a finite-dimensional $\mathbf{F_q}$-vector space of polynomials (namely those of degree $< k$) at a finite set of points in $\mathbf{F_q}$. These points can be thought of as lying on a line and Goppa's new idea was to replace this line by a curve C of genus g. We then pick n distinct points $P_1, \ldots, P_n$ on C and a divisor D whose coefficients at these P_i are zero (we call D disjoint from the P_i). Generalizing the Reed-Solomon code, we consider the code $\{(\phi(P_1), \ldots, \phi(P_n)) \mid \phi \in \mathrm{L}(D)\} \subseteq \mathbf{F_q^n}$.

This is an $[n,k,d]$-code over $\mathbf{F_q}$ with $k \geq \deg(D) + 1 - g$ (a bound coming from the Riemann-Roch theorem) and $d \geq n - \deg(D)$. It follows that the rate $R = k/n$ and relative minimum distance $\delta = d/n$ are related by $R + \delta > 1 - g/n$. Since we want both R and δ large, we want n/g large, in other words curves with small genus but lots of points. This brings us back to the question addressed by Niederreiter and Xing above.

By studying the behavior in towers of certain well-known curves in number theory called modular curves, Tsfasman, Vladut, and Zink [15] showed that for q a square greater than or equal to 49, the asymptotic behavior of the algebraic-geometry codes these curves produced was unusually good. In particular, they were the first to beat the Gilbert-Varshamov bound, which means that their rate R and relative minimum distance δ satisfied $R > 1 - H_q(\delta)$, where $H_q(x) = x\log_q(q-1) - x\log_q(x) - (1-x)\log_q(1-x)$ (for $0 < x < 1 - 1/q$) is the Hilbert entropy function. Many believe that this cannot be beaten for $q = 2$, in which case the function $\alpha_2(\delta)$ introduced earlier would be $1 - H_2(\delta)$.

References

1. Ahlswede, R., Cai, N., Li, S.-Y.R., Yeung, R.W.: Network information flow, EEE Trans. Inform. Theory, **46**, 1204–1216 (2000)
2. Alamouti, S.M.: A simple transmit diversity technique for wireless communications". IEEE Journal on Selected Areas in Communications **16**, 14511458
3. Goppa, V.D.: Geometry and codes, Kluwer (1988)
4. Hochwald, B.M., Sweldens, W.: Differential unitary space time modulation, IEEE Trans. Inform. Theory, **46**, 543-564 (2000)
5. Hughes, B.L.: Differential space-time modulation, IEEE Trans. Inform. Theory, **46**, 2567–2578 (2000)
6. Koetter, R., Médard, M.: An algebraic approach to network coding, IEEE/ACM Trans. Networking, **11**, 782–796 (2003)
7. Niederreiter,H.: Random number generation and quasi-Monte Carlo methods, CBMS-NSF Regional Conference Series in Applied Math., Vol. 63, Soc. Industr. Applied Math., Philadelphia (1992)
8. Niederreiter,H., Xing, C.: Rational points on curves over finite fields - theory and applications, London Mathematical Society Lecture Notes Series 285 (2001)
9. Nitinawarat, S., Boston, N.: A complete analysis of space-time group codes, in Proceedings of the 43rd Annual Allerton. Conference on Communication, Control, and Computing
10. Paskov, S. H., Traub, J. F.: Faster evaluation of financial derivatives, J. Portfolio Management, **22**, 113-120 (1995)
11. Shokrollahi, A.: Computing the performance of unitary space-time group codes from their character table, IEEE Trans. Inform. Theory, **48**, 1355 –1371 (2002)
12. Shokrollahi, A., Hassibi, B., Hochwald, B., Sweldens, W.: Representation theory for high-rate multiple-antenna code design, IEEE Trans. Inform. Theory, **47** , 2335–2367 (2001)
13. Stichtenoth, H.: Algebraic Function Fields and Codes, Springer-Verlag (1993)
14. Sturmfels, B., Shokrollahi, A., Woodward, C.: Packing unitary matrices and multiple antennae networks, in preparation.
15. Tsfasman,M.A., Vladut,S.G., Zink, T.: Modular curves, Shimura curves, and Goppa codes, better than Varshamov-Gilbert bound, Math. Nachr. **109**, 21–28 (1982)
16. Zassenhaus, H.: Uber endliche Fastkorper, Abh. Math. Sem. Univ. Hamburg., **11**, 187–220 (1936)

Chapter 5
Emerging Applications to Signal Processing

Abstract In this chapter we discuss some new applications of algebra to signal processing. The first way in which this arises is via the transfer function of a filter. This leads to questions about polynomials and rational functions, which can sometimes be solved by ideas from algebraic geometry, often with better results than through numerical methods. The second application of algebra is to image processing. We already saw how three-dimensional rotations can be simply described by quaternions – now we look at an application in face recognition.

5.1 Filters

A (discrete-time) *filter* is a system that takes an input sequence (signal) (x_n) and turns it into an output sequence (y_n), usually with some unwanted component removed. A simple example is smoothing out a signal by taking short-term averages, say $y_n = (x_{n-2} + x_{n-1} + x_n)/3$. This is what is called a low-pass filter because it allows low but not high frequency sinusoids to pass through. The effect of this filter can be studied by attaching generating functions $X(z) = \sum x_n z^{-n}, Y(z) = \sum y_n z^{-n}$. The effect of a linear recurrence relation then translates into multiplying the generating function by the z-transform of its impulse response or its *transfer function*.

Theorem 5.1. *If the filter is described by* $y_n + a_1 y_{n-1} + \ldots + a_p y_{n-p} = b_0 x_n + b_1 x_{n-1} + \ldots + b_q x_{n-q}$, *then* $Y(z) = X(z)H(z)$, *where* $H(z) = \sum_{k=0}^{q} b_k z^{-k} / (1 + \sum_{k=1}^{p} a_k z^{-k})$.

Proof. Set $a_0 = 1, a_k = 0$ for k outside the range $[0,p]$, and $b_k = 0$ for k outside the range $[0,q]$. Then $(\sum b_k z^{-k})(\sum x_n z^{-n}) = \sum_{k,n} b_k x_n z^{-k-n} = \sum_m (\sum_{k+n=m} b_k x_n) z^{-m} = \sum_m (\sum_{k+n=m} a_k y_n) z^{-m} = \sum_{k,n} a_k y_n z^{-k-n} = (\sum a_k z^{-k})(\sum y_n z^{-n})$

5.1.1 Makhoul's Conjecture

An example of this is given by the famous Makhoul's conjecture [7]. Suppose the recurrence relation has the form $y_n + a_1 y_{n-1} + ... + a_p y_{n-p} = a_p x_n + a_{p-1} x_{n-1} + ... + a_1 x_{n-p+1} + x_{n-p}$. Then $H(z) = z^{-p} A(z^{-1})/A(z)$ is the transfer function of an (all-pass) filter. If the input signal is $x_n = \delta_n$, which means that $x_0 = 1$ but all other x_n are zero, then the output signal y_n has the form $A_1 \alpha_1^n + ... + A_p \alpha_p^n$, if $\alpha_1, ..., \alpha_p$ are the (distinct) complex zeros of $A(z)$.

If these roots all have absolute value less than 1, then the output signal is stable, meaning that $y_n \to 0$ as $n \to \infty$. This implies that $|y_n|$ takes on a maximum value for some n and Makhoul conjectured that this n is at most $2p-1$. This was proven for $p = 2$ [2] but shown to be false for $p = 6$ [10]. The true bound is now thought to be of the order of $p \log p$, but this problem is open.

5.1.2 Pipelining

Another example concerns the pipelining of filters [3]. Suppose $H(z) = B(z)/A(z)$, where $A(z)$ and $B(z)$ are polynomials in z^{-1}. The goal of pipelining is to obtain better throughput at the expense of latency by modifying the architecture without changing the transfer function. Mathematically this amounts to multiplying the numerator and denominator of $H(z)$ by the same polynomial $D(z)$ in z^{-1} such that the first few coefficients of the new denominator $A(z)D(z)$ are zero or powers of 2. We also need $D(z)$ to be stable. The largest absolute value of a pole of $D(z)$ will be called its pole radius, which should be as small as possible to ensure robust stability.

Example 5.1. A certain Butterworth filter has $A(z) = 1 - 2.3797z^{-1} + 2.9104z^{-2} - 2.0551z^{-3} + 0.8779z^{-4} - 0.2099z^{-5} + 0.0218z^{-6}$. We want to find $D(z)$ of the form $1 + d_1 z^{-1} + d_2 z^{-2} + ... + d_L z^{-L}$ such that $A(z)D(z) = 1 + c_1 z^{-1} + ... + c_{L+6} z^{-L-6}$ with $c_1, ..., c_6$ equal to 0 or a power of 2.

Now $D(z) = A(z)D(z)/A(z) = 1 + (2.3797 + c_1)z^{-1} + (2.7526 + 2.7397c_1 + c_2)z^{-2} + ...$ Every "good" choice of $c_1, ..., c_6$ determines $d_1, ..., d_6$ and then the question becomes the following.

Given $d_1, ..., d_6$ and $L \geq 6$, find $D(z) = 1 + d_1 z^{-1} + ... + d_L z^{-L}$ with as small a pole radius as possible. We will call $D(z)$ the pole radius minimizer of degree L of $1 + d_1 z^{-1} + ... + d_6 z^{-6}$. If it is stable (i.e. has pole radius < 1), we say that $1 + d_1 z^{-1} + ... + d_6 z^{-6}$ has a stable extension.

Theorem 5.2. $1 + az^{-1}$ *has a stable extension of degree L if and only if* $|a| < L$.

Theorem 5.3. $1 + az^{-1} + bz^{-2}$ *has a stable extension of degree* 2 *(i.e. is stable already) if* (a, b) *lies in the stability triangle, defined as the interior of the triangle bounded by the lines* $b = 1, b = a - 1, b = -a - 1$.

What about a stable extension of degree 3? What is its pole radius minimizer? This is where algebraic geometry has the advantage over numerical methods. In comparison, consider the old problem of finding two positive numbers with sum 1 and product as small as possible. Numerical methods, such as Newton's method, can be used to approach the solution. We know, however, by calculus that the answer is exactly when the two numbers are equal. A similar phenomenon arises here.

Theorem 5.4. *Given any real numbers a,b, the degree 3 pole radius minimizer of $1+az^{-1}+bz^{-2}$ is $1+az^{-1}+bz^{-2}+cz^{-3}$ where $c \in \{ab, b/a^3, ab/2-a^3/8, (a(9b-2a^2)\pm 2(a^2-3b)^{3/2})/27\}$.*

Many more examples can be found [3]. The point is that the pole radius minimizers correspond to relations among the roots α,β,γ, namely $\alpha=-\beta, \alpha\beta=\gamma^2, \alpha+\beta=\gamma, \alpha=\beta$. In general:

Theorem 5.5. *Given $a_1,...,a_M \in \mathbf{R}$ and $L \geq M$, there exists a computable variety $V(M,L)$ inside $\mathbf{R^L}$ of dimension M such that the degree L pole radius minimizer of $1+a_1z^{-1}+...+a_Mz^{-M}$ has $(a_1,...,a_L) \in V(M,L)$.*

5.1.3 The Belgian Chocolate Problem

Here is another example of a problem where numerical methods have obtained reasonable solutions but an algebraic geometry approach betters them. The problem was posed by Blondel in his thesis [1] to illustrate how hard problems in stabilization of plants that appear simple can be. In this case, stability of a polynomial will mean that all its zeros lie in the left half-plane.

Suppose $0<\delta<1$. Can we find stable polynomials $p(s), q(s)$, and $r(s)$ such that $(*): (s^2-2\delta s+1)p(s)+(s^2-1)q(s)=r(s)$?

This arises if you start with the unstable "plant" $P(s)=(s^2-1)/(s^2-2\delta s+1)$ and add a feedback loop with a controller $C(s)=q(s)/p(s)$, then the combined transfer function has denominator $1+P(s)C(s)$. It is desirable that the numerator and denominator of $C(s)$ are stable and the point of stabilization is for the denominator of the combined transfer function to be stable. This yields the problem as asked.

If $\delta=1$, then the left side and hence $r(s)$ has a root at $s=1$ and the question is how close to 1 can δ be taken. Blondel [1] offered 1kg of Belgian chocolate for the solution to this and 1kg for a solution when $\delta=0.9$. He claimed in his thesis that there is no solution for $\delta>0.9999800002$.

Let us stratify the problem by imposing the condition that the degrees of $p(s)$ and $q(s)$ are at most n. For example, for $n=6$, the best published solution [5] is $\delta=0.96292177890276$ and (where we have normalized $p(s)$ to have leading coefficient 1)

$$p(s)=z^6+1.9267063716832z^5+2.7125040416507z^4+3.2971535543909z^3$$

$$+2.4444879197567z^2 + 1.4102519904753z + 0.7321239653705,$$

$$q(s) = 1.1928111395529z^2 + 0.0002957682513z + 0.7321239653705.$$

In fact, these exact polynomials are not stable as given because of rounding issues after normalizing. This is the whole point. The numeral optimization methods of [5] are limited purely by the number of decimal places they were working to. Increasing the precision would allow better solutions. We now indicate what those solutions are tending towards.

Looking closer at the above solution, we note that it has several approximately repeated roots and is very close to the form $p(s) = (s^2 + 2\delta s + 1)(s^2 + A)^2, q(s) = k(s^2 + B), r(s) = s^8$. Plugging this back into $(*)$ and equating coefficients of s^0, s^2, s^4, s^6 yields:

$$A^2 - kB = 0$$

$$-4\delta^2 A^2 + kB - k + 2A^2 + 2A = 0$$

$$-8\delta^2 A + k + A^2 + 4A + 1 = 0$$

$$-4\delta^2 + 2A + 2 = 0$$

These four equations in four variables can be solved explicitly (or else numerically to any desired precision). For example, the last equation gives $A = 2\delta^2 - 1$. The third equation then gives $k = 12\delta^4 - 12\delta^2 + 2$ and then the first equation gives $B = (2\delta^2 - 1)^2/(12\delta^4 - 12\delta^2 + 2)$. Plugging these into the second equation then yields the equation $16\delta^6 - 16\delta^4 + 1 = 0$, which yields δ and hence k, A, B.

Taking δ to be the largest real root of $16\delta^6 - 16\delta^4 + 1 = 0$, namely $\delta = 0.962973955961027$, gives positive A and B and the solution

$$p(s) = (s^2 + 2\delta s + 1)(s^2 + 2\delta^2 - 1)^2 =$$

$$s^6 + 2\delta s^5 + (4\delta^2 - 1)s^4 + (8\delta^3 - 4\delta)s^3 +$$

$$(4\delta^4 - 1)s^2 + (8\delta^5 - 8\delta^3 + 2\delta)s + (4\delta^4 - 4\delta^2 + 1),$$

$$q(s) = (12\delta^4 - 12\delta^2 + 2)s^2 + (4\delta^4 - 4\delta^2 + 1).$$

These polynomials are not stable because they have roots on the imaginary axis, but it is possible to deform their coefficients slightly, at the expense of lowering δ by an arbitrarily small amount, such that the deformed polynomials are stable. Plugging in numerical values gives polynomials close to the one above obtained by Chang and Sahinidis [5]. They did not obtain better values of δ because those solutions can only seen to be stable by working to high precision. For more examples and the currently largest solution for δ, namely 0.976462, see [4].

It is interesting to note a connection between this problem and the *abc* problem in polynomials. Mason and Stothers [8], [11] showed that if three relatively prime polynomials $a(s), b(s), c(s)$ satisfy $a(s) + b(s) = c(s)$, then the largest degree among the three is less than the number of distinct zeros of $a(s)b(s)c(s)$. This implies an analog of Fermat's Last Theorem for polynomials since if $a(s)^n + b(s)^n = c(s)^n$, it follows that each of $n \deg a, n \deg b, n \deg c$ is less than the number of distinct zeros of

$a(s)^n b(s)^n c(s)^n$, which is less than $\deg(abc)$. Adding these three inequalities yields $n\deg(abc) < 3\deg(abc)$ and so $n < 3$.

The critical cases always come close to this bound, if we set $a(s) = (s^2 - 2\delta s + 1)p(s), b(s) = (s^2 - 1)q(s)$, and $c(s) = r(s)$. In fact, in each case, the bound is satisfied with a difference of only 2.

The extremal examples (where the bound is achieved with a difference of just 1) correspond to covers of the projective line $\mathbf{P}^1$ minus three points and so to Grothendieck's dessins d'enfants. Our examples are related to covers of $\mathbf{P}^1$ minus 4 points, which are in turn classified by points of a certain Hurwitz moduli scheme. One systematic way to produce many critical cases might be to investigate the corresponding points on the Hurwitz scheme.

5.2 Image Processing

One important area of signal processing is image processing. We have seen how three-dimensional rotations can be compactly described by quaternions of norm 1. One big problem in image processing is face recognition and one of the many challenges to overcome is to recognize faces that present themselves in different orientations. This could be overcome by storing faces at different angles (expensive on storage) or by computing a rotation to bring the face into a normalized position (expensive on computation), but an elegant solution that apparently matches what humans really do is to use invariants. The idea is that, however the face presents itself, the various curvatures of the nose for instance will not change. Since derivatives are sensitive to noise and because curvature involves second derivatives, this method does not work well in practice. For more detailed discussion of this section, see Olver's survey article [9].

We formalize matters as follows. If G is a group, then we say that G acts on a set M if there is a given map $\phi : G \times M \to M$. We usually denote $\phi(g,m)$ by $g \circ m$ indicating the result of the effect of g acting on m. If G and M have a topology, then we usually insist that the action is a continuous map. The *orbit* of m is the set $O_m = \{g \circ m \mid g \in G\}$. An *invariant function* is $I : M \to \mathbf{R}$ such that $I(g \circ m) = I(m)$. In other words, I is constant on each orbit.

Example 5.2. Let $G = SO_2$, the group of two-dimensional rotations about the origin. Let $M = \mathbf{R}^2$. The orbit of $m \in M$ is the circle with center at the origin and radius $|m|$. One obvious invariant is $I : M \to \mathbf{R}$ defined by $I(m) = |m|$ and a little thought shows that every invariant is a function of this I. We call I the *fundamental invariant.*

For a general G and M, we seek a set of fundamental invariants $I_1, ..., I_k$ meaning that every invariant is $H(I_1, ..., I_k)$ for some function H and none of the I_i is redundant. There is a general method when G is a Lie group (a smooth manifold that is also a group) acting smoothly on a manifold M. It is called Cartan's method of moving frames.

Definition 5.1. A *moving frame* is a smooth map $\rho : M \to G$ such that $\rho(g \circ m) = g \circ \rho(m)$ holds.

A moving frame exists in a neighborhood of a point m if and only if G acts freely and regularly there. Free action means that the isotropy subgroup $G_m := \{g \in G \mid g \circ m = m\}$ is trivial. Regular means that O_m intersect a neighborhood of m is connected. In the above example, a moving frame is defined (except at the origin) by mapping m to g_θ where θ is the argument of m.

Another useful notion is canonical form, which means picking one element from each orbit. In other words, we have a set $K \subseteq M$ such that $|K \cap O_m| = 1$ for every $m \in M$. For the above example, K can be taken to be the positive real axis.

Suppose now that G is an r-dimensional Lie group acting on an m-dimensional manifold M with $r < m$ freely and regularly. If we pick coordinates for M such that $K = \{(z_1, \ldots, z_m) \mid z_1 = c_1, \ldots, z_r = c_r\}$, suppose that $g^{-1} \circ z = (w_1(g,z), \ldots, w_m(g,z))$. Then the fundamental invariants are $w_{r+1}(\rho(z), z), \ldots, w_m(\rho(z), z)$.

For our running example, this amounts to converting to polar coordinates, in which $K = \{(r, \theta) \mid \theta = 0\}$ and the other variable r gives the fundamental invariant.

In cases when the action of G is not free and regular, there are ways to enlarge the manifold by introducing derivatives (differential invariants) or by taking products of M with itself (joint invariants). For instance, given a smooth curve C in the plane, we might associate to each point $(\kappa, d\kappa/ds)$, which will trace out a curve called the signature curve of C. If we rotate and/or translate C to get an equivalent curve C', they have the same signature curve. Amazingly, the converse is true.

Theorem 5.6. *Two planar curves are equivalent if and only if they have the same signature curve.*

As for surfaces in $\mathbf{R}^3$ (which includes faces we want to recognize), there is a corresponding signature surface in $\mathbf{R}^6$, with the property that two surfaces are equivalent if and only if they have the same signature surface. Thus, mathematically, the recognition problem is solved by simply having a database (or gallery) of signature surfaces and, given a face, comparing its signature surface with those in the gallery. This does not account for the elasticity in a face, but there is another critical issue that arises from this not being a perfectly mathematical matter, namely that the signature surface is unduly sensitive to noise.

In response to this, we developed a new kind of invariant based on integrals rather than derivatives. These have greater robustness and as a result our team came second nationally in the Face Recognition Grand Challenge in a 3D competition. For further information on this work, see [6].

References

1. Blondel, V.D.: Simultaneous stabilization of linear systems. Lecture Notes in Control and Information Sciences **191**, Springer-Verlag (1994)
2. Boston, N.: Makhoul's conjecture for $p = 2$. In: IEEE International Conference on Acoustics, Speech, and Signal Processing (2001)
3. Boston, N.: Pipelined IIR filter architecture using pole-radius minimization. VLSI Signal Processing **39**, 323–331 (2005)
4. Boston, N.: On the Belgian Chocolate Problem and output feedback stabilization: efficacy of algebraic methods. To be submitted.
5. Chang, Y.J., Sahinidis, N.V.: Global optimization in stabilizing controller design. Journal of Global Optimization **38**, 509–526 (2007)
6. Lin, W.-Y., Boston, N., Hu, Y.H.: Summation invariant and Its applications to shape recognition. IEEE International Conference on Acoustics, Speech, and Signal Processing (2005)
7. Makhoul, J.: Conjectures on the peaks of all-pass signals. IEEE Signal Processing Magazine **17**, 8–11 (2000)
8. Mason, R.C.: Diophantine equations over function fields. London Mathematical Society Lecture Note Series, **96** (1984)
9. Olver, P.J.: A survey of moving frames. In: Computer Algebra and Geometric Algebra with Applications, Lecture Notes in Computer Science **3519**, 105–138 (2005)
10. Rajagopal,R., Wenzel, L.: Peak locations in all-pass signals - the makhoul conjecture challenge. In: Proceedings of the International Conference on Acoustics, Speech, and Signal Processing (ICASSP) 2001
11. Stothers, W.W.: Polynomial identities and Hauptmoduln. Quart. J. Math., Oxf. II. Ser. **32**, 349–370 (1981)

Glossary

Alamouti scheme The first space-time code discovered.

Cyclic code A code invariant under cyclic permutation of coordinates.

Diversity product A measure of how good a space-time code is.

Full diversity A space-time code with nonzero diversity product.

Fundamental invariants A minimal set of invariant functions that produce all other invariant functions.

Golden code An example of a perfect space-time code.

Halton sequence A particular example of low discrepancy sequence.

Hamming code A code whose parity check matrix has every vector as a column.

Hamming distance The number of places in which two vectors differ.

Hamming weight The number of nonzero entries in a vector.

Invariant function A function on a set that is unchanged by a group acting.

Linear feedback shift register Producer of sequences by linear regression.

Low discrepancy sequence A sequence whose star discrepancy has a particularly small order of magnitude.

Moving frame A special kind of smooth map from a manifold to a group acting on it.

Niederreiter-Xing sequence Particularly good examples of low discrepancy sequences.

Orbit All images of one element under a group acting on a set.

Parity check A linear equation satisfied by all vectors in a code.

Perfect code A space-time code with all the desirable properties.

Pole radius Largest absolute value of a pole of a transfer function.

Quadratic residue code A special kind of cyclic code.

Quaternion algebra A division ring, 4-dimensional over its base field, constructed similarly to Hamilton's quaternions.

Quaternions Elements of a division ring discovered by Hamilton in 1843.

Signature curve/surface A curve/surface that is invariant under group action on curves/surfaces.

Space-time codes Collections of well-spaced matrices used for coding both in space and time.

Stable Has all its roots in the unit circle or left half-plane, depending on context.

Star discrepancy A measure of even distribution of sequences of points in a hypercube.

Strict rank criterion Where all nonzero matrices in the ambient set have full rank.

Transfer function The z-transform of the impulse response of a filter.